Epigenetics
&
Genetic
Happiness

Epigenetics
&
Genetic
Happiness

HOW TO INCREASE
YOUR WELL-BEING

by the author of the bestselling
Happiness Genes

James D. Baird, Ph.D.

for further information
on the author, his works,
and his blog see
www.drjamesbaird.com

Naperville, Il
2019

Disclaimer: This book is not designed to replace medical or psychiatric treatment for a serious health condition. Please seek professional help if you have questions about your physical or psychological health.

Contents

Introduction

These days, there's a lot of interest in your genes. Multiple services offer to trace your genetic background for you and tell you where the genes you've inherited came from. Medical companies race to improve DNA testing that will indicate your vulnerability to physical disorders and identify mutations in your genes that may cause illness or disease.

There's also a growing interest in the study of mechanisms that can switch genes on and off. Although you can't control your genes directly, you can have some control over which genes are dormant and which are active. That science is called *epigenetics*.

The word *epigenetics* was coined back in the 1940s to refer to the influence of genetic processes on human development. It took a while for the topic to attract much attention, but in 2006, over 2,500 articles about epigenetics were published.

Then interest grew. By 2010—the year that I brought out my book *Happiness Genes: Unlock the Positive Potential Hidden in Your DNA*—that number jumped up to over thirteen thousand publications. Unlike many of those, my books have always been designed to make topics clear to everyday readers.

In 2012, the *International Journal of Epidemiology* declared "Epigenetics: the next big thing."[1] By then, I had published *Obesity Genes and their Epigenetic Modifiers,* offering applications for the new discoveries.

Interest continued to grow, and in 2013, there were over seventeen thousand[2] publications on topics related to epigenetics. In 2015, I published *Behavioral Genes: Why We Do What We Do and How to Change* based on the abundant research at that time.

Now new research findings regarding human lives and cultures have led to this new book, *Epigenetics and Genetic Happiness: How*

1 Discussed by Shah Ebrahim.
2 According to Carrie Deans and Keith A. Maggert in *Genetics.*

to Increase Your Well-Being. We'll look into ways that you can actually bring about changes in your biological mechanisms to shape the sense of well-being that's so basic to your personal happiness.

NOTE: Informal footnotes throughout this book direct you to complete source information in a section following the text.

1.
The Components of Well-Being

Basically, we humans want to be happy. That's a natural, healthy, and widespread human desire. Because it's such a strong desire, we give and get a lot of advice on the matter. Authors and psychologists and advertising executives all try to tell us what will make us happy. Stories in books, films, and television shows give us popular versions of the path to living happily ever after.

If all that advice worked, we wouldn't have so many people still searching for happiness or complaining that it hasn't been provided to them. We wouldn't have collected so many things that don't make us happy at all. That's a terrible waste of time, of resources, and of human lives. We need to redefine, reorganize, rethink the question. But most of the time we can't define what happiness is, much less how to achieve the happiness we want.

An article in *Psychology Today* described happiness as an "elusive state," and that's how it seems for many people. It may be easier to say what happiness is not, or how you don't attain it.

HAPPINESS IS NOT . . .

According to that *Psychology Today* report,

> Research shows that happiness is not the result of bouncing from one joy to the next; achieving happiness typically involves times of considerable discomfort.[3]

Happiness isn't even an automatic reward for attaining goals. In 2018, nearly one-fourth

---

3 From "Happiness" in *Psychology Today.*

of undergraduates signed up for a university class designed to teach students how to lead a happier life. It became the most popular class ever taught in Yale University's 316-year history. The professor, Dr. Laurie Santos, pointed to one big reason why:

The things Yale undergraduates often connect with life satisfaction—a high grade, a prestigious internship, a good paying job—do not increase happiness at all.[4]

In fact, a study published in *Proceedings of the National Academy of Sciences* challenged the idea that what we often refer to as happiness is even a good thing:

Happiness without meaning characterizes a relatively shallow, self-absorbed or even selfish life, in which things go well, needs and desire are easily satisfied, and difficult or taxing entanglements are avoided. . . .[5]

---

4 "Yale's Most Popular Class Ever: Happiness," *New York Times,* Jan. 26, 2018.
5 This comment is quoted by Emily Esfahani Smith in her article in *The Atlantic.*

In an article for *The Atlantic,* author Emily Esfahani Smith reported the reseachers conclusion,

> People who are happy but have little-to-no sense of meaning in their lives have the same gene expression patterns as people who are enduring chronic adversity. [6]

In other words, that kind of shallow "happiness" isn't good for your body. The study found more beneficial patterns of gene expression associated with meaningfulness in life. (In this book you'll find more about that term *gene expression*: what it is, how it affects us, and how to use it.)

SO WHAT DO WE MEAN BY HAPPINESS?

I'm certainly not going to try to define what *things* will make you happy. Happiness definitely means different things to different people. Instead, we can begin by considering the entire question in more useful terms.

---

6 From Emily Esfahani Smith's article in *The Atlantic.*

We can generally agree that happiness is a positive emotional experience. Most of us recognize that we can't sustain it at a high level all of the time. (That's probably a good thing. If happiness were constant, how could we tell happiness from unhappiness?)

Back in 1996, after a career as an engineer and inventor of high-tech equipment, I sold my business and retired to work and write on the topic of happiness as the purpose of life. I'm still finding that a rich and rewarding topic— one might even say it's a meaningful contribution to a happy life.

Well-being

In a lot of definitions of happiness, you'll see the word *well-being*. Someone might state that happiness is essential to a sense of well-being, or vice versa—that well-being is essential to happiness.

The term *well-being* implies a sense of satisfaction and contentment, even of personal fulfillment.

So let's work with this definition:

> Happiness is a mental or emotional state of well-being characterized by positive or good emotions, ranging from satisfaction to joy.

Of course, things like health, security, and environment contribute to your sense of well-being. But there's more to well-being than that. Some secure and healthy people don't seem to be able to find happiness at all.

Some factors that affect your well-being might be beyond your control, but the truth is that a lot is actually under our own control. That doesn't mean well-being is gained by getting more things. (That can just become a matter of accumulating debris.)

You also don't need to spend your life battering against things that won't change. You can put the bad things in perspective and still live with a general sense of well-being

There are three important things that affect your ability to experience well-being. Understanding these is important, because there are some you can't change—but there are some that you can.

The first is genetic happiness, which is inherited from your parents. The second emanates from epigenetic (environmental) mechanisms, most of which you have some control over by epigenetic therapies. The third is life itself. Some of what you experience, such as childhood events, are beyond your control, but other adventures are yours to shape for yourself.

GENETICS, THE INHERITED FUNDAMENTALS

Genes can be thought of as the blueprints that provide the design for the human body and for how it develops. The word *genome*—a combination of the words *gene* and *chromosome*—refers to the genetic information of any organism. The human genome is often called the "map" of our DNA.

Genes you've inherited might or might not be active in your own makeup. As you know, you don't exhibit absolutely every trait of your parents and other ancestors. When a gene is active, that's called gene expression. Some genes express themselves and others don't.

You can't change your genes, but you can change their expression. That's because biochemicals called markers *on* genes (therefore *epi-genetic)* have control over what is or is not activated. And you can change those markers that control your genes.

These days, many people in our culture have taken a new interest in their genetic backgrounds. Simple tests give us fascinating information about where our ancestors came from and what mix of backgrounds makes up our own being.

We know now that what we inherit isn't limited to characteristics such as height and hair color. Your inheritance also shapes your behavior.

Your genetics are inherited, and that inheritance makes up approximately 30–40 percent of your personality. But take note: that's not even half of who you are.

Even though you might inherit a basic approach to life, that will naturally vary a bit. Your attitude can move up or down, responding to positive or negative events. Those changes in attitude generally take place for a relatively short period of time, and then you return to a familiar range of feelings called a set point. (More about that in chapter 2.)

Epigenetics—what you can change

In recent years, you might have noticed the sharp rise of interest in an important field called *epigenetics*. It's about surprising ways that you can make changes in your well-being.

Your genes are with you for life, but they don't always control your life. Discoveries in behavioral science have shown that our genes can sometimes be turned on or off. Certain circumstances will cause genes to become dormant. Other circumstances cause them to be active, to "express" themselves.

Scientific studies have shown that physical changes in organisms can be caused by modification of the way that genes are expressed, without any changes in the genetic code itself. That's exciting news. It explains that we control part of our well-being at a very basic level.

So epigenetics isn't just a topic for researchers tucked away in universities and other laboratories. It's important to know about in everyday life. You can't change your genes. But the way your genes express themselves can be changed. In fact, that expression *is* changed by life itself. And some of those changes can be inherited by future generations.

CULTURE MAY CHANGE YOU

Part of our well-being is affected by the conditions in which we live. That doesn't just mean the obvious. Of course, if you live in famine or other horrifying conditions, your well-being will be affected. But there have to be more subtle reasons why, even among well-off nations, there is a consistent inequality in perceived happiness.

Global polls and studies indicate that the happiness in the United States has been declining for at least a decade. We're not near the top among developed nations. We come in about fourteenth among countries with modern economies.

As an article in *Time* put it:

Another year, another report saying that the Nordic countries are the happiest in the world.[7]

In chapter 4, we'll look at the reasons why the United States doesn't turn up at the top of lists of happiest countries. What's different

---

7 From the "Happiness Report" in *Time*.

between the culture we live in here and those of other nations?

Taking charge of your own well-being

To live in a state of well-being, some things are helpful to recognize, and there are things you can change. That's what this book is about.

Your epigenetic makeup does change, because markers on your genes are affected by your environment. And at least some of those circumstances are within your control. Besides that, you can use specific epigenetic therapies to support your own well-being.

One website on the topic of epigenetics compares human life to a long movie. The script is your DNA. Epigenetics is like the director, who can tweak the script.[8]

You are in charge of that director.

---

8 From "A Super Brief and Basic Explanation of Epigenetics for Total Beginners" on the website *What Is Epigenetics?*

2.
Genetic Happiness

Perhaps you've been told that your disposition is just like your mom's or dad's, or that you're like some relative further removed—possibly even someone you've never met. Just as you might have inherited your appearance from some ancestor, you also might have inherited some important aspects of your *self.* That's because some parts of our personalities and behaviors are based in our genes.

It actually doesn't take a lot of genetic difference to give each of us our own basic personality. Ninety-nine percent of every person's genes are identical to every other person's anyhow. But that's not remarkable when you consider that we actually share about 98 percent of our genes with chimpanzees, our closest relatives in the animal kingdom.

A very small amount of genetic individuality makes you who you are. Less than 1 percent of our genes is different from those of other people—and that small percentage makes each of us look and behave in our own unique way. Studying the effects of that 1 percent has occupied scientists in various disciplines over many years.

Now new data from one of the largest studies ever published tells us a lot about human happiness and what we can do about it.[9]

FINDING THE HAPPINESS GENE

A huge study that involved over 190 researchers in 140 research centers in seventeen countries has analyzed data from hundreds of thousands of people. The researchers were looking for the genetic variations associated with our sense of well-being and also with other traits—depression and neuroticism.[10] They reported

---

9 Okbay, Aysu, and others in *Nature News*.
10 See "Happiness Genes Located for the First Time."

their findings in the journal *Nature*, and their work has also been widely covered in other sources.

The team commented on the state of research as well as on the results:

- An estimated 350 million people suffer from depression worldwide.
- Current understanding of the role of genetics in character traits is patchy.
- This most recent study has pinpointed variants in genes associated with depressive traits and neuroticism.
- Unpacking the individual role of genes and environment is challenging.[11]

As these researchers noted, your genetic contributions to your personal perspective aren't easy to determine and analyze. Although this team made no claims for finding final answers, they did come up with useful information. Alexis Frazier-Wood, assistant professor of pediatrics and nutrition at Baylor College of Medicine, Houston, reported that the team

11 Catharine Paddock, "Happiness Gene: Does It Exist?"

"found three genetic variants associated with subjective well-being—how happy a person thinks or feels about his or her life."[12]

In other words, the research team discovered what has been called the "happiness gene."[13] So that means that a certain amount of your happiness does depend on what you're born with. (However, that doesn't mean you're stuck with it if you don't like what you got. This book includes epigenetic therapies for managing your own genetic inheritance.)

Of course, that isn't an end to the research. Professor Meike Bartels, VU University Amsterdam, commented,

> This study is both a milestone and a new beginning: A milestone because we are now certain that there is a genetic aspect to happiness and a new beginning because the three variants that we know are involved account for only a small fraction of the differences

12 Catharine Paddock, "Happiness Gene: Does It Exist?"
13 See "Happiness Genes Located for the First Time."

between human beings. We expect that many variants will play a part.[14]

THE HERITABILITY OF HAPPINESS

Geneticists use the term *heritability* to define the variations in personality and behavior that are caused by differences in genes. It turns out that even our sense of well-being has a genetic component. In other words, happiness can be heritable.

Behavioral Genetics is the field of study that examines the role of genetics in animal and human activities. It's an interdisciplinary field, drawing on biology, genetics, epigenetics, psychology, and statistics.

Behavioral geneticists study our inheritance of our behaviors.

A lot of research in this field has been based on observations of twins identified through the Netherlands Twin Register, which was founded in 1987 in Amsterdam. The NTR follows information on twins and other multiples and their parents, siblings, and spouses. Over 175,000

---

14 Catharine Paddock, "Happiness Gene"

people are registered in all. Data collected on these people is used to study hereditary effects on various factors, such as personality, growth, development, and risk factors for disease.

This information is often gathered through observations of identical twins in different environments and in comparisons between the two different types of twins.

Monozygotic twins are perfectly identical genetically because they come from the same fertilized egg. In contrast, dizygotic twins share only 50 percent of their genes—just as ordinary brothers and sisters do. By comparing the personalities of pairs of monozygotic and dizygotic twins, scientists can estimate the heritability of personality traits.

Researchers are particularly interested in comparing twins raised together by the same parents to twins raised separately by different families. That makes it easier to separate genetics from environmental influences on personality.

If identical twins raised separately since birth have similar personalities, then those similarities are probably due to the fact that they have the same genes.

Those studies have found that some degree of certain personality traits can be traced to genetics. For example, research on tens of thousands of twins shows that the trait of extroversion due to inheritance might be as high as 50–60 percent.

For neuroticism, research shows that the difference between emotionally stable people and neurotic people (those who tend to be moody, emotional, anxious, and prone to stress) is a little less than half—30–50 percent can be traced to genetics.

ON THE EUDAEMONIC SEESAW

Our sense of well-being depends on what kind of happiness we're trying to find. Most people pursue common happiness, technically referred to as *hedonic*, or extrinsic happiness. This is happiness without meaning. It comes from external gratifications, such as material acquisitions, relationships, life situations, and shallow sources of pleasure that can be easily satisfied without consideration for others.

This common happiness is transient because it depends on external circumstances, which always change.

Natural happiness is intrinsic. It isn't something you can buy; it is something you give. The Greek word for it, *eudaemonic*, is still used by social scientists today. This is happiness *with* meaning, and it is the result of genetic motivations.

This kind of happiness often springs from a belief in a God or universal spirit and depends upon giving of yourself through good deeds and compassion. It is not derived from temporary pleasures, but from altruistic acts.

The competitive survival genes that sustained us during our prehistoric days on the African savannas have first priority when life seems threatening. Our selfless, cooperative genes were developed later, during the agricultural period about ten thousand years ago. When the tribal culture of our ancestors melted into agricultural communities, cooperation was necessary to plan, plant, and harvest crops. In other words, the tribe that works together stays together.

Some of those ancestors learned to pursue intrinsic eudaemonic goals. As a result, they had a higher survival rate and passed along their genes at a higher rate. That's why we do see compassion in the world today.

In spite of that adaptation for cooperation, sometimes our older genes take over—those that we inherited from ancestors who weren't big on cooperation. That's why we see conflict and prejudice as countries, groups, and individuals compete for resources and power.

The tilt back and forth continues for individuals and within nations.

EPIGENETICS AND OUR ANCESTORS

Even our ancient ancestors experienced epigenetic influences. Little did they know that diet, life experiences, climate, toxic environments, and competing tribes modified the proteins that affected their health and even their own behaviors.

Today, our environment is totally different. Where the early human was on his own to survive and protect his family, now we have legal systems, the police, and healthcare.

Our modern culture gives us the opportunity to modify those old behaviors that were effective in their time but that don't always work to our benefit today. A case in point is aggression. Without aggression our prehistoric ancestors, like other animals, would not have survived. They had to protect themselves and their turf.

Today aggression isn't usually helpful, and our culture imposes severe penalties against it. However, unimpeded by those cultural legal restrictions, aggression again rises to ancient standards in countries at war.

We now know that our inheritance can be modified and that those changes lead to greater happiness. So how can we harness the power of our minds and create positive changes for ourselves? You'll find specific techniques in the final chapters of this book.

Research has shown that obtaining intrinsic goals improves emotional well-being, but achieving extrinsic goals has no such effect. For example, people who spend more time and effort on positive relationships have a higher level of well-being. Other intrinsically rewarding activities include social volunteering, hobbies, and serving charities. Conversely, making money is hedonic and not intrinsically rewarding. It is not valuable for its own sake but is to be used for some other purpose.

In fact, eudaemonic happiness is an evolutionary adaptation that promotes health. A study published in the *Proceedings of the National Academy of Sciences* in 2013 found that happiness affects the human genome.[15] Those people with higher levels of eudaemonic happiness had lower levels of inflammation and stronger expression of their antibody and antiviral genes.

In contrast, those with high levels of hedonic happiness had high inflammation and weak expression of antibody and antiviral genes.

You and your set point

An important finding of the study confirmed that there is a genetic set point of happiness. That means each of us has an "average" level of happiness,[16] the range of which remains fairly constant throughout life. That's called a set point. For a short time, your set point might change due to events, but it tends to return to

15 Discussed by E.E. Smith in "Meaning Is Healthier Than Happiness."
16 From "The Set-Point Theory of Happiness" on the *Changingminds* website.

the same range. Even positive events, such as career advancement, money, and marriage, only have temporary effects on our set points.

A certain amount of our usual range of well-being is inherited from our parents, and it's also affected by early life experiences. Although a set point isn't easily shifted, some research suggests certain things that can do it:

A few events—chief among them the unexpected death of a child and repeated bouts of unemployment—that seem to reduce our ability to be happy permanently. Yet some studies also suggest that we can fix our happiness set point permanently higher—by helping others.[17]

Research and real life have both demonstrated that more acquisitions don't increase happiness. Even self-esteem—a healthy amount of which is necessary for happiness—doesn't promote happiness at all when it's increased to the point of narcissism. Compared to achieving goals like career advancement, money, and marriage, compassionate action is what has been found to make us happier in the long

---

17 From "How to Reset Your Happiness Set Point" in *Psychology Today.*

run.[18] This effect is universal to all people in all cultures and independent of changing circumstances. It's hardwired into our brains.[19]

---

18 Discussed in "How to Reset Your Happiness Set Point" in *Psychology Today.*
19 You'll find more about this in my book *Happiness Genes.*

3.
Epigenetics and Emotions

As we discussed in chapter 1, your sense of well-being is composed of three distinct elements. The first is genetic happiness, which is inherited from your parents. The second kind of well-being stems from epigenetic mechanisms, and the third element is made up of life experiences. How much control you have over each of those elements varies.

For example, events that have already happened—such as childhood experiences—may have a powerful effect on you, but the events themselves are beyond your ability to change. Other life events can be managed in present time or planned for in advance. Larger-scale cultural situations probably aren't within your direct control, although you can adjust your exposure to some of them. In any case, you do have some control over how any of these events affect you.

You might think in a similar way about genes.

Genes don't make decisions about what they do or whether they're turned on or off. An article in *Discover* magazine put it this way:

> A human liver cell contains the same DNA as a brain cell, yet somehow it knows to code only those proteins needed for the functioning of the liver.[20]

An *epigenome* (the word means "above the genome") contains chemical compounds that tell genes what to do. They're also called "gene markers." These chemical messages can turn on or turn off specific genetic information, controlling how a gene shapes the organism. The *Discover* article suggests that we "think

---

20 Ethan Watters, "DNA Is Not Destiny: The New Science of Epigenetics."

of the epigenome as a complex software code, capable of inducing the DNA hardware to manufacture an impressive variety of proteins, cell types, and individuals."

Scientists once thought that the patterns of the human epigenome were set during early fetal development. More recent discoveries show that the epigenome can and does change during your entire lifetime. Alterations are made in response to your environment, which includes your surroundings, experiences, diet, and personal behavior. These changes take place without affecting your DNA, but in some cases they're heritable, meaning they can be passed on to offspring.

EPIGENETIC MARKERS

In the early 1940s, the British embryologist Conrad Waddington began using the term *epigenetics* to describe interactions between genes and certain biochemicals.

You can't control what genes you inherit, but it turns out that you do have some control over how those genes affect you. That's what the new science of epigenetics has revealed, and there are ways to put that science to practical, everyday use. Your epigenetic markers can be rewritten—which means that you can modify the instructions that your genes receive. Epigenetic therapies help you reshape the way your genetic inheritance affects you.

So if genes are like a blueprint to build and operate a human, then you can think of proteins as the contractor that does the work of building the organism. You can change those proteins with epigenetic signals—and those signals include beliefs and perceptions. That's because your perception of any given thing, at any given moment, can influence your brain chemistry. It influences the chemistry of your blood, which in turn influences your cells and controls the expression of your genes. In other words, your thoughts and perceptions have a direct and significant effect on the genes and their proteins in your cells.

That's why there has been so much excitement about the science of epigenetics. It shows us how to control damage caused by familiar and unhelpful human responses.

Basically, achieving epigenetic satisfaction is a matter of balancing your emotions.

EMOTIONS

If you ask someone how they're doing, they're likely to reply with words that describe their feelings—happy, joyful, sad, afraid, frustrated. Because emotions are often considered very personal and private, a person you ask might not tell you anything at all about how they really feel. Even then, since we all share similar emotions, you can often guess how people feel from their behavior.

If someone displays a particular emotion most of the time, we might call that their disposition or personality.

Early Greek philosophers had opinions about emotions and their impact on life. Aristotle thought they were important to achieving moral excellence, but the group called Stoics thought that emotions got in the way of reason and were detrimental to a virtuous life. Ancient Chinese sages also believed that too much emotion led to an unfortunate imbalance in the life force, or *qi*.

In the late nineteenth century, the naturalist Charles Darwin concluded that emotions have evolved along with our bodies. Based on his theory of evolution, he believed that emotions serve a purpose and evolve through natural selection in the same way that physical traits evolve. Darwin expected similar emotions to show up in people from all different kinds of backgrounds—and they do.

These days, professionals in many fields study human emotions. Psychiatry, nursing, psychology, neuroscience, linguistics, education, sociology, anthropology, communications, economics, philosophy, and the arts all contribute theories about what emotions mean and, in some cases, how they can be manipulated.

WHAT ARE EMOTIONS?

Emotions are defined by recent research as a system of thoughts, feelings, motives, and bodily reactions that are interconnected. They're evolutionary adaptations that alert us to situations that might be important because they threaten our survival. In addition, positive emotions stimulate our well-being.

There's a difference between emotions and moods. Displays of emotion may only last a few seconds, while moods can last for hours, days, or even longer. A positive or negative mood usually draws related thoughts and memories along with it. Moods give flavor and meaning to the events of our lives.

An emotion is almost always a response to an event—something that happens or might happen—that could affect your own well-being. Sometimes that event is real, and sometimes it's imagined. In either case, emotions are the driving force behind what you do about it. For better or worse, emotions motivate us, just like physical sensations such as hunger or pain. They can propel us to action or freeze us in inaction. Although they can be very intense at times, they're usually brief.

Our emotional reactions can be influenced by hormones and neurotransmitters, such as dopamine, noradrenaline, serotonin, oxytocin, cortisol, and GABA (gamma-Aminobutyric acid). They're also influenced by our personalities, our present mood, and our way of thinking. For example, the difference between intuitive and critical thinking is the same as the difference between emotional thinking versus factual thinking.

For another example, a person with an extroverted personality is likely to react less to a potentially menacing glance then a person with a neurotic personality. That's because extroverts enjoy interacting with people, while neurotics usually see the negative side of everything that happens. Extroverts and neurotics will have different interpretations of the same glance and different expectations of what might follow.

For all of us, our perceptions of whether a situation will be a danger, a benefit, or of no concern to our well-being are shaped by the way we think and, consequently, by their effect on our emotions. Do we use intuitive thinking and respond emotionally, without facts, or do we use critical thinking and analyze the situation in keeping with actual evidence?

Your emotional reaction to a situation can also vary according to the environment at the time. That same potentially menacing glance might seem different if you're in a crowded and well-lit room rather than a dim and desolate alley.

WHERE DID EMOTIONS COME FROM?

If you tell someone to change an emotion— "stop worrying," for example—you're likely to be met with a blank stare. The reason for that is rooted far back in human history.

Evolutionary psychologists have built on Darwin's ideas. They've integrated theories from research that connects emotions with the activation of certain areas of the brain. Like Darwin, those psychologists believe that emotions evolved to protect our prehistoric ancestors. Early humans lived on African savannas, where rustling in the bushes might just be just wind—or might be a hungry tiger. Exaggerated emotions could have saved lives back in those days.

Like other animals, our prehistoric ancestors reacted to their perceptions of threats and benefits. However, other animals couldn't project an image of themselves facing the same situation at some other time, so they didn't waste energy worrying about the past or fearing the future. Human beings developed the self-awareness to consider what might yet happen. Then, they could be fearful of possibilities as well as of immediate realities.

Of course, the alarm our ancestors experienced was generally in response to a real hazard. Although we're not likely to be under such constant threat today, our minds still generate defensive emotions. It doesn't take much of an imagined threat to get our motors racing.

Consider road rage, for example, or the way some of us respond to negative criticism. At one time or another, most of us worry about the past and fear the future, whether or not there's a real danger out there.

THE POWER OF EMOTIONS

Emotions can certainly focus your attention. They're related to specific urges, and they demand a response. If you don't respond, those emotions won't let you alone. (Think obsessive-compulsive syndrome.)

If you believe you have a problem but try to ignore it, your emotions can be like a dog chewing on a bone, gnawing at you until you take action and resolve the matter. When that nagging negative emotion disappears, you know that the problem has been taken care of.

Emotions also play a key role in changing our biology so that we are better able to respond to threats. A well-known example is the fight-or-flight syndrome that increases our heart rate, respiration, and blood pressure. An emotion is preparing the body for possible jeopardy, and that kicks in whether the danger is real or not.

Of course, emotions don't just alert us to potential threats. Current research suggests that they influence most of the decisions that we humans make. They can drive us to seek relationships, to create art, to excel athletically, to engage socially, or to compete politically. Emotions can also lead us into destructive behaviors, such as seeking unbridled power or revenge.

It's important for us to understand how these strong emotional forces shape our lives. Let's look at one in particular that haunts us all from time to time.

Ads for computer dating sites like to ask, "Are you lonely?" Apparently enough people answer "yes" to keep a lot of those operations going strong.

Anybody—anywhere, at any age, at any income level—can experience loneliness. It's especially prevalent in older people who feel isolated. Some therapists believe that we feel lonely because when we were very young, we needed someone to take care of us, and our subconscious mind still worries over that.

Evolutionary psychologists point out that we are social animals who rely on each other to survive. They say that the feeling of loneliness originally developed to alert early humans to stay inside the crowd. Being too close to the perimeter of the group increased the risk of becoming prey. But even without any tigers lurking in the tall grass, loneliness can kill.

Researchers at Brigham Young University investigated the correlation between social relationships and mortality. Their mega analysis of 148 studies covered more than three hundred thousand participants. According to an article in *The Wall Street Journal,* the study found that loneliness is a strong predictor of early death. Loneliness is harder on your health than obesity or a sedentary lifestyle. It's just as bad for you as alcoholism or smoking fifteen cigarettes a day.[21]

---

21 From Elizabeth Bernstein, "When Being Alone Turns Into Loneliness, There Are Ways to Fight Back."

THE ROLE OF THE UNCONSCIOUS

Our unconscious mind is an evolutionary adaptation that evolved to defend our ancestors from perceived threats to their survival and well-being. It is totally defensive. It only thinks intuitively, a reflex commonly referred to as a "gut reaction."

It considers everything in relation to the past and develops habits to defend against past mistakes. It defends against perceived threats by keeping you worried and on guard.

Our unconscious mind made this adaptation for good reason. A quick gut reaction is required if you're facing a "flight-or-fight" situation. Time is of the essence. That emotional thinking can save your life or prevent you from doing something stupid.

However, those worries are then projected into threatening events in an illusionary future. That's why those worries about the past and fears of the future constantly bounce around in our heads.

Later, different thinking skills were needed to navigate the world. Critical thinking replaced emotional thinking with facts. That transition allows us to see that some threats are perceived rather than real, and this eliminates a lot of unnecessary worry.

(The final chapters in this book include therapies for using critical thinking rather than intuitive responses.)

We're still sometimes driven by the warnings and trepidations of something that's like a voice in our heads. It's a random stream of thoughts that can take virtual control of our thinking and put us in a state referred to as being on autopilot. That ancient voice clamors on, always searching for potential dangers, both here and now and in all our future possibilities. Our emotions flow randomly, usually full of fears.

How epigenetics works

It turns out that the new science of epigenetics has a lot to say about emotions, too. Epigenetic therapies are designed to help you, and they do work.

Here's why:

The main thing that destroys our ability to lead a satisfactory life is emotional suffering.

Simply put, emotional suffering is caused by the worries and fears of our unconscious mind.

For example, meditation works because our minds can only think of one thing at a time. In mindful meditation, you become aware of all your surroundings at the moment. Since your mind is in the moment, you cannot worry, because worry is about past perceived threats. (For more about meditative therapies, see the chapter on mindfulness.)

The voice in our heads is the voice of our unconscious minds.

If the voice in our heads stops, so do the worries.

That voice can be turned off.

4.
Cultural Well-Being

According to a major poll taken every year of people of various nations, certain countries shuffle in and out of the top spot for being the happiest. The United States has never made the top ten. In fact, the United States has been continuously declining in measures of happiness for the past decade.

The most recent unpleasant news comes from *The World Happiness Report 2018*, a survey of the state of global happiness. The rankings are compiled from results provided by Gallup World Poll surveys taken over several years. The report publishes a ranking of levels of happiness around the world and notes important changes. The United States always turns up far down that list compared to other developed nations.

In 2018, Finland was at the top of the happiest countries list. Norway and Denmark have also held that position, and the other Nordic countries are always high on the list. The Netherlands and Switzerland have also been at the top.

In the 2018 report, the United States turned up in eighteenth place—four levels lower than the previous year. In spite of economic improvement in recent years, happiness levels in the United States keep falling.

It's not that we give no attention to the question of happiness. In fact, we're constantly being told what will make us happy. Consider the design of television ads that focus on the satisfaction that having x or y is supposed to bring you. We're pummeled with the notion that happiness can be bought—if you're smart enough to gain the financial resources you need to acquire it. And, of course, those recommended purchases are supposed to show everyone else just how successful you've become at being happy.

But it that's true, why does the United States keep turning up so far down that list of happiest countries? Surely a lot of Americans can buy at least as much stuff as most citizens of other countries. Even so, U.S. citizens experience high levels of depression and mental disturbances.

LOW RATINGS ON HEALTH CARE

In addition to its overall rankings, *The World Happiness Report 2018* also focuses on several specific problems related to health and well-being: obesity, opioid use, and depression. In the United States, all three of those problems have grown faster and spread wider than in most otherwise similar countries.

Worse, we seem to be doing less than others to counteract that terrible trend. In the *Happiness Report*, author Jeffrey Sachs[22] wrote:

America's public health . . . has improved much less than in most other high-income countries, and in

22 A Professor at Columbia University, Sachs is known as one of the world's leading experts on economic development and the fight against poverty.

recent years is experiencing an outright
decline. The U.S. life expectancy actually fell by 0.1 years from 2014 to 2015,
and then by another 0.1 years from
2015 to 2016.[23]

That fall in life expectancy at birth is
most unusual for high-income countries in
peacetime.

According to a 2017 evaluation made by the
think tank Commonwealth Fund, U.S. healthcare was rated worst among eleven developed
nations:

In addition to ranking last or close to
last in access, administrative efficiency,
equity and health care outcomes, the
U.S. was found to spend the most
money on health care.[24]

IDENTIFYING IMPORTANT DIFFERENCES

The World Happiness Report 2018 ranked
156 countries by their happiness levels. So

---

23 Sachs, "Chapter 7: America's Health Crisis and
the Easterlin Paradox."
24 See Ryan Bort's article on the U.S. healthcare
system in *Newsweek*.

coming in at number eighteen means that the U.S. is still near the top of the list. The top countries have strengths in common, and they all provide benefits for their populations.

All the top countries tend to have high values for all six of the key variables that have been found to support well-being: income, healthy life expectancy, social support, freedom, trust and generosity.[25]

Clearly, U.S. citizens are not less happy than those who live in truly difficult situations. The question is, why do Americans report being unhappier than the citizens of nations that are comparable or even lower in resources? The problem isn't that we can't buy enough stuff, as if that were going to make us happy. It might be that we have to buy more basic things than the citizens of those countries do.

The difference between the cultures of happier countries and the U.S. is mainly how much the government satisfies the needs of most of the population. In general, the governments of the more highly ranked countries satisfy more of the economic, social, and healthcare needs of most of their populations.

---

25 *World Happiness Report 2018.*

Those Nordic countries that stay high on the happiness list—Denmark, Finland, Norway, and Sweden—are not all alike, but they do share some features worthy of attention. Norway, Sweden, and Denmark are constitutional monarchies in which the king or queen holds a mostly symbolic position. Finland has never had a monarch. Decision making in those nations is in the hands of an elected legislature. Those various governments work within the structure of capitalism to promote economic security for their citizens.

In these countries, taxes are high, but services are very high. Citizens get free higher education, medical services, and care for the elderly. The governments provide protection for workers, support for those out of work, and help finding a new job when needed. Poverty rates are very low.

The Nordic systems aren't error-free, but the citizens of those countries do keep reporting high levels of happiness.

RELIGIONS AND CULTURES

"Sweden is a highly secular nation, and Swedes appear to see little connection between religiosity and happiness,"[26] according to a Swedish government website. Only eight percent of all Swedes regularly attend religious services. Religion is still visible in traditional rituals or ceremonies, such as christenings, marriages, and funerals and in religious holidays, but it's not a regular part of daily life. Sweden is number nine in the *World Happiness Report.*

As reported in Gallup polls, the other Nordic countries rank slightly higher in religion. The United Kingdom and the Netherlands also rank high in populations that do not consider themselves religious, followed by Germany, Switzerland, Spain, and Austria. About half of the French population is not religious. In Israel, it's 65 percent.[27] The highest numbers of all in atheist citizens is in China, number eight-six on the happiness report, and Hong Kong, number seventy-one.

A Pew Research Center report found that richer countries tend to be less religious than poorer nations. However, income doesn't seem

---

26 From the website Religion in Sweden.
27 From "World's Least Religious Nations-or the Atheist Countries....." on Pakistan Defence.

to account for their positions on the happiness scale.

HIGH INCOMES AND LOW HAPPINESS

A chapter of *The World Happiness Report 2018,* written by Jeffrey Sachs, included a section called "restoring American happiness," which addressed some issues in the United States:

> The central paradox of the modern American economy, as identified by Richard Easterlin (1964, 2016), is this: income per person has increased roughly three times since 1960, but measured happiness has not risen. The situation has gotten worse in recent years: per capita GDP is still rising, but happiness is now actually falling... America's crisis is, in short, a social crisis, not an economic crisis. [28]

One simple experiment brought this home to researchers in 2017.

28 Sachs, "Restoring America's Happiness" in the World Happiness Report 2017.

Stamped and addressed envelopes were dropped in public areas (sidewalks, shopping malls, phone booths), to see whether people pick them up and put them in a mailbox. This is a measure of helping behavior among strangers.[29]

On a repeated use of that test, U.S. residents dropped 10 percent in helping behavior between 2001 and 2011, although helping among residents of neighboring Canada didn't change.[30]

Sachs identifies those results as due to an underlying problem in U.S. culture—a growing "us versus them" attitude.

ZERO-SUM INTERACTIONS

In an era when resources were limited to whatever was available in a particular environment, our ancestors had to work out ways to deal with shortages. If things were scarce on ancient African savannas, inhabitants had to take control of what was available or move on

---

29 Sachs, "Restoring America's Happiness" in the World Happiness Report 2017.
30 See Hampton, "Why is Helping Behavior Decliningin the United States But Not in Canada?

if they were to survive. Competition developed. For some, the view became "either I get it or you get it, but not both of us."

The assumption that one can only gain if another loses has been called "zero sum" in game theory and economic theory. In an example given by author Robert Wright, we are told this:

> In zero-sum games, the fortunes
> of the players are inversely related.
> In tennis, in chess, in boxing, one
> contestant's gain is the other's loss. In
> non-zero-sum games, one player's gain
> needn't be bad news for the other(s).
> Indeed, in highly non-zero sum games
> the player's interests overlap entirely. In
> 1970, when the three Apollo 13 astro-
> nauts were trying to figure out how
> to get their stranded spaceship back
> to earth, they were playing an utterly
> non-zero-sum game, because the
> outcome would be either equally good
> for all of them or equally bad. (It was
> equally good.)[31]

31 Robert Wright in *NonZero: The Logic of Human Destiny,* p 5.

Wright admits that things in the real world aren't often so clear-cut. But he's describing the difference between expecting to beat the other out for scarce resources or working together to create more resources for all. It could be that if we're looking through a zero-sum lens, the notion of working together for equal benefits isn't even visible as an option.

THE TWO-PARTY PROBLEM

If two equal groups of people are competing for resources, perhaps they fall easily into an us-or-them frame of mind. If the competitors belong to three or more groups, they might be more likely to make concessions and work cooperatively with some others in order to share or even increase resources.

Back when the United States began, the founders had misgivings about the dividing power of factions. George Washington wasn't in favor of political parties at all. In his farewell speech, our first president warned that founding parties based on "geographical discriminations" would lead the country back to something like a monarchy:

The alternate domination of one faction over another, sharpened by the spirit of revenge, natural to party dissension, which in different ages and countries has perpetrated the most horrid enormities, is itself a frightful despotism. But this leads at length to a more formal and permanent despotism. The disorders and miseries, which result, gradually incline the minds of men to seek security and repose in the absolute power of an individual.[32]

In our two-party form of government, a small percentage of our population has managed to control most of the resources, making it unrealistic to consider the need-fulfillment programs of the Nordic nations. Our system results in increased stress and anxiety for an ever-larger percentage of citizens.

DEALING WITH CONFLICT

Conflict in the world has always been about competition over resources, and, consequently,

---

32 From George Washington's Farewell Address.

equalization of resources results in the least conflict. The unfortunate conclusion is that grossly unequal resources in the United States will continue to result in increasing conflict and unhappiness.

However, we can deal with that at an individual level. Since conflict over resources is a core genetic response, epigenetic therapies can change genetic expression and greatly reduce that stress. Stress reduction will result in more emotional satisfaction for the person who practices it.

5.
How We Think

Sometime back in ancient human history, we began to think about something more than survival. It took a very long time for that to happen.

According to historian Yuval Noah Harari, creatures identifiable as humans started making stone tools about two-and-a-half million years ago.[33] As far as we can tell, those minds weren't conscious of anything more complicated than getting through from day to day—finding food and water, procreating, battling off predators and perhaps other humans. They did very little else for about two million years.

---

33 Yuval Harari in *Sapiens: A Brief History of Humankind.*

Our own species, *Homo sapiens,* finally put in an appearance about two hundred thousand years ago. By then, humans had gained control of fire for daily use. Some scientists believe that the consumption of cooked food helped change the species into creatures more like humans we know today. Other scientists have other explanations. But in any case, we're still talking about slow progress over a very long period of time.

Some sixty thousand years ago, our ancestors were still operating with a basic form of thinking that served them well. (Even those early ancestors did attain a certain level of intelligence. Before fifty thousand years ago, they were smart enough to build boats to reach Australia.)

STRENGTHS OF THE ANCIENT MIND

The quick reactions of early human minds were still well suited for human survival on the African savanna even some sixty thousand years ago. In those days, there was no time for critical thinking. Stopping to consider all aspects of a problem and all possible solutions would have been a disaster for them—and probably would have provided an easy meal for predators.

That ancient mind continuously played back its genetic programs, instincts, and life experiences. Therefore, it was an unconscious mind. Back then, those thoughts weren't just idle chatter; they generated emotions that focused attention to real problems. Those early human brains were hardwired with survival behaviors. Their emotions provided a basic survival mechanism, promoting flight on the chance that flight might be essential.

But a change took place.

Somewhere along its grinding process, evolution came up with an upgrade. This new design included a physical increase in the size of the forebrain, which enabled new ways of thinking.

Yuval Harari believes a change in consciousness began to happen when humans reached a point when they could think beyond what they already knew—when they could create fiction. He places that at about fifty thousand years ago.[34]

---

34 Yuval Harari in *Sapiens: A Brief History of Humankind.*

This "add-on" mind was our conscious mind; it was designed for a different kind of thinking. However, this new mind didn't come with quick-start instructions. It took vast psychic effort and life experience for human beings to learn how to even turn on and use this conscious mind.

At some point, we were finally no longer completely controlled by our original-equipment mind. It has been suggested that the breakthrough came when we human beings learned how to think more about the present and ignore thoughts tied to the past. Our primeval mind can't operate when we're "in the moment" because it's so connected to the past.

A change in human thinking is certainly connected with lifestyle changes brought about by the agricultural revolution, about 10,000 BC. That period, called the Neolithic, was when the development of agriculture moved our ancestors from tribal to community living.

Before then, when our tribal ancestors came across food, they ate, for they never knew when they would have their next meal. This changed with the agricultural movement. Having a reliable food supply was the motivation for change, but little did those early humans realize what would be involved.

The agricultural movement initiated the evolution of a conscious mind to supplement the ancient unconscious mind—a major step in evolution.

To produce a more dependable food supply, our ancestors had to learn to work with others, even those who were not from the same tribe. They had to plan ahead for a future harvest and wait to see that anticipated reward for their efforts. This new agricultural life required social cooperation rather than tribal competition.

In the process of becoming good farmers, our ancestors needed a leader to direct group efforts. One reasonable speculation is that this leader's commands soon became ingrained in the followers' brains. And when that leader died, his voice was still in the followers' minds, perhaps giving rise to the first conception of a God or universal spirit.

Psychologist Julian Jaynes believed that humans were directed by actual voices in our heads as late as 1000 BC (remnants of the sounds of those tribal leaders).[35] Although recent research finds that biologically improbable, his work supports the idea that the consciousness we know is a recent development, especially when you consider the great length of human history.

THE ANCIENT MIND IS WITH US STILL

The primeval part of our minds is still functional. In fact, our older unconscious mind has considerably more "computing power" than our newer conscious mind, and it's almost always in attendance. So our unconscious mind still grabs most of our attention, making it hard for us to use the power of the newly evolved mind.

Others call that ancient mind "gut thinking."

Gut thinking is human nature, and it's taken advantage of by promoters who profit by misleading people. That includes much of the advertising field, various media, politicians, and others who want to sell products, services, and anything else that can play to our intuitive mode of thinking.

---

35 Discussed in Julian Jaynes, *The Origin of Consciousness in the Breakdown of the Bicameral Mind.*

With dedicated practice, we can overcome that block. The procedures are simple and easily available. There are countless methods of meditation, mind control, stress management, and other mental exercises used to develop more consciousness. In this book, the recommended procedures for mindfulness can be found in the chapter by that name. This isn't the only mindfulness procedure, but its epigenetic effects are well understood, giving it more support than other practices.

MIND AND BODY

Whatever the connection is between mind and body, it is an evolutionary adaptation. The connection permits the mind to influence the body in beneficial ways, such as in alternative and complementary therapies.

Consider the placebo. When a fake treatment—a sugar pill or a saline solution—improves a patient's condition just because the patient expects it to work, we call that the "placebo effect." It demonstrates the power of the mind.

About 40 percent of the time, a believing patient who receives a placebo will experience an expected improvement. Simply put, your belief can make you well at least 40 percent of the time. Research has shown that if you believe in the provider, and the provider believes in the effectiveness of the treatment, that 40 percent can grow to 80 percent.

The bottom line is that we can use the mind-body connection to change our behaviors by changing our beliefs and perceptions. The mechanism for doing this is described by that science we've been talking about—epigenetics.

The unconscious mind is still powerful. Sometimes it needs to be quieted down so that we can put the power of our conscious minds to use.

THE POWER OF THE CONSCIOUS MIND

Since our conscious mind evolved much later than our unconscious mind, it's ignorant of primitive survival behavior. It understands the benefits of cooperation, altruism, compassion, and creativity. It would appear that the designer of evolution understands that love of neighbor is a far more reliable survival plan than stealing his Toyota or bombing his house. Unfortunately, our level of conscious thinking constitutes a very small part of our overall thinking. However, there is some progress, as exemplified by humanitarian efforts and acts of compassion that shine through here and there.

Your conscious mind is creative, effective at future planning, and not subject to your past survival needs and selfish acts. It promotes the state of "we rather than me." In that mental state, altruism and compassion preside, and genetic happiness and joy can reign.

Simply put, the solution to the problems presented by your unconscious mind is to become more conscious. The most effective way of doing this is through the mental epigenetic therapies presented in the final chapters of this book.

6.
Feeling Good

If you ask yourself what makes you happy, what answers do you come up with? Do you think of your relationships . . . or the way you live . . . or things you own? Do you have something in mind that you're sure would make you happy if only you could attain it?

Sadly, too many of us spend too much of our lives pursuing popular myths about happiness. And they are myths. As I noted in my 2010 book Happiness Genes, an array of seductive media constantly bombards us with messages about what will make us happy,but evidence suggests that the pursuit of happiness according to the following four assumptions invariably leads to disappointment.

They are:

1. MORE is better
2. NEW is better
3. BIGGER is better.
4. FASTER is better.[36]

Although acquiring things as recommended in these myths may have a short-term positive affect, we soon return to our genetic happiness set point—that personal average level of happiness that I described in chapter 2. (See the subhead "You and your set point.")

Because each individual has a tendency to return to their ordinary range of happiness, having more than others doesn't keep us feeling better for very long. No matter what we acquire, we only feel good about it for a short time.

Here's an observation about set points by clinical psychologist Robert Puff:

Next time you're flying coach, perform this simple experiment: As you walk

---

36 *Happiness Genes,* p. 29.

past passengers seated in their big fluffy business class seats, check out the expressions on their faces. Does every single one of them have a big smile as they embrace flying in luxury? As a professional observer of human behavior, I've done this experiment countless times, and the answer I come up with every time is the same: They don't look any happier than the other passengers.[37]

When we return to our set point, some of us try to jog our happiness level up again by constantly acquiring new things. Because that only works for a short time, we can be trapped in a never-ending scramble for new stuff that won't make us happy for very long.

Having a set point that is rather low on a happiness scale can even lead to a state of desperation. It drives some people into constant effort—even criminal activity—just to regain that temporary feel-good state.

---

37 Robert Puff, "Your Set Point for Happiness" in *Psychology Today.*

THE CHEMISTRY OF HAPPINESS

The truth is that what we think of as happiness is already built into us. It evolved as our species developed. And it's basically chemical.

Four main neurochemicals, hormones, and neurotransmitters are responsible for feelings of happiness. They're generated in our own brains. So you can forget all those glitzy ads. You don't need to rush out and acquire new things in order to enjoy your life. You've already got what you need for feeling good.

Mindfulness teacher Kaia Roman reminds us why it's a good thing that happiness is a matter of chemistry.

> This is actually great news. It means even when circumstances, possessions, or people in our lives aren't exactly as we'd like them to be, there are simple ways we can increase our happy brain chemicals and alter our moods.[38]

Even when we aren't happy with the ways our circumstances ebb and flow, there are

38 Kaia Roman, "The Brain Chemicals That Make You Happy."

epigenetic mechanisms we can use to increase our happy brain chemicals. Better yet, some foods and activities help produce more than one of these important substances.

The magic four brain chemicals are endorphins, serotonin, dopamine, and oxytocin. They are neurotransmitters, called chemical messengers, that transmit signals from one nerve cell to another.

Epigenetic therapies in the final chapters of this book will be useful in increasing your own production and activation of these natural chemicals that contribute to your sense of well-being. Let's look at what they are and how they work.

ENDORPHINS

Produced by the central nervous system, endorphins create a positive feeling that's similar to the effect of morphine (but without the risks of addiction). You might have heard someone refer to "an endorphin rush" or a "runner's high." That's those natural chemicals kicking in.

Endorphins are activated by pain, and they rush to the rescue to help us deal with physical discomfort. They interact with our brains to reduce our perception of pain and to provide a sedative effect. That's why some athletes actually enjoy participating in very demanding sports.

Fortunately, endorphins are also activated by milder exercise. If you're not practicing extreme athletics, you can still increase the endorphins in your system. In fact, endorphins are released by just about any type of physical activity (including sex). In one experiment described by Kaia Roman, clinically depressed subjects found improvement after they simply walked on a treadmill for just thirty minutes, ten days in a row.

No drugs, no shiny new possessions, just walking.

Studies have shown that endorphins are also activated by acupuncture, massage therapy, and the practice of meditation. Some foods, such as chocolate and hot chili peppers, also promote endorphin release.[39]

---

39 See "Endorphins: Natural Pain and Stress Fighters" on MedicineNet.

SEROTONIN

This one is actually known as "the happiness chemical." Like endorphins, serotonin transmits messages between nerve cells, contributing to your sense of well-being.

According to research reports, low levels of serotonin and clinical depression go hand in hand, although it's not clear whether low serotonin causes depression or is caused by it.[40] Feeling low, experiencing poor memory, being irritable and anxious—all are associated with low serotonin.

Although some recreational drugs release serotonin, those can lead to unwelcome symptoms lasting for days and even long-term nerve damage. Fortunately, you can activate serotonin with a range of perfectly ordinary activities.

Low sunlight can cause a drop in serotonin, which can trigger a low mood. For some people this is just a seasonal problem, occurring when we naturally have less exposure to sunlight. However, many people suffer from low serotonin year round. Fortunately, there are also other methods of boosting this neurotransmitter.

---

40 See the article by James McIntosh in *Medical News Today.*

Exercise, which, of course, has other benefits too, helps produce serotonin. Spending time in an outdoor setting has also been shown to help produce a good mood. So make it a point to get outside and move around.

Foods heavy in tryptophan also produce serotonin. Yes, this means chocolate—but that's not all. Tryptophan is an amino acid that your body needs but can't make for itself. It has to come from your diet. Fortunately, tryptophan is abundant in foods such as oats, dried dates, milk, yogurt, cottage cheese, red meat, eggs, fish, poultry, sesame, chickpeas, almonds, sunflower seeds, pumpkin seeds, buckwheat, spirulina, and peanuts.

Even though foods high in simple carbohydrates (pastas, potatoes, bread, etc.) do increase tryptophan, you're better off using the ones with more complex carbohydrates (such as leafy vegetables, squashes, sweet potatoes, fruits, beans).

You can feel the calming effect of serotonin pretty quickly, usually within a half hour of eating the right foods. (For more on maintaining a healthy diet, see chapter 14.)

Even your thoughts can produce serotonin—if they're happy thoughts. In fact, feelings of gratitude have been shown to produce both serotonin and dopamine. (Also see the breathing exercises and meditation techniques in the final chapters of this book.)

DOPAMINE

The neurotransmitter dopamine is responsible for that jolt of pleasure you feel when you get something right—when you hit a target or accomplish a task. It's often called the "chemical of reward." Dopamine stimulates us to seek more of the activity or substance that brings pleasure.

Dopamine is cognitive enhancing, helping with focus and attention. It controls the flow of information to the frontal lobes from other areas of the brain. A shortage of dopamine means a decline in functions such as memory, attention, and problem-solving.[41]

Certain foods (like chocolate), sex, and some drugs of abuse stimulate the release of dopamine. In fact, the concentration of dopamine is the reason that some recreational drugs,

---

41 From Ananya Mandal, "Dopamine Functions."

such as cocaine and amphetamines, are addictive. Fortunately, we do have other methods of increasing these happiness chemicals.

For example, volunteering has been shown to increase dopamine. According to psychology professor Tracey Alloway,

> Giving of your time or volunteering can release the same feel-good sensation as eating chocolate or a candy bar. And that's because brain scans show a surge of dopamine when we give or volunteer our time. Researchers call this "the helper's high."[42]

According to some research, even *thoughts* of loving kindness can bring on that dopamine high.

OXYTOCIN

This one is a hormone that also acts as a neurotransmitter. It might be familiar to mothers because it's produced during pregnancy

---

42 Tracey Alloway, "Give to Feel Good" in *Psychology Today.*

and breastfeeding. A 2007 study reported that high levels of oxytocin during their first trimester of pregnancy has a positive effect on women later bonding with their babies.[43]

Oxytocin has been called the "love hormone." It's activated during sex, and it also appears that cuddling a loved one, even hugging a pet, can increase oxytocin levels. People in the early stages of romantic attraction have been found to have increased levels of oxytocin.[44]

Oxytocin reduces depression and anxiety and contributes to relaxation. It's associated with feelings of empathy and trust, so it can be important in building relationships. Clinical psychologist Carol Rinkleib Ellison commented that "It's like a hormone of attachment, you might say."[45]

Oxytocin actually stimulates the production of dopamine and serotonin. So when this hormone is released into your system, it's also helping other beneficial chemicals kick in.

---

43 Maureen Salamon, "11 Interesting Effects of Oxytocin."
44 Markus MacGill, "Oxytocin: The Love Hormone?"
45 Maureen Salamon, "11 Interesting Effects"

So how do you raise your oxytocin levels? Many studies report that touching and hugging someone, or even a pet, will increase that hormone level. Researcher Paul J. Zak recommends eight hugs a day. He also suggests giving another person your complete attention, giving a gift, sharing a meal, soaking in a hot tub, or telling someone that you love them.[46]

Meditation is also often recommended for increasing oxytocin, so you can use the techniques in later chapters for this benefit too.

---

46 Paul J. Zak, "The Top 10 Ways to Boost Good Feelings" in *Psychology Today.*

7.
Feeling Worthy

Your sense of self-worth is often called self-esteem. At various times, esteem for yourself has been declared a basic human need, an inalienable right, and a reflection of social status. Some measure of self-esteem is generally considered important for everyone to have.

If you search "self-help" in Amazon books, then click the "Self-Esteem" subhead, you get tens of thousands of results. In a *Psychology Today* article, Heidi Grant Halvorson noted that most books on the topic "aim to not only tell you why your self-esteem might be low, but to show you how to get your hands on some more of it."[47]

---

47 Heidi Grant Halvorson, "Forget Self-Esteem" in *Psychology Today.*

Experiencing the ups and downs of self-esteem involves beliefs such as "I am worthy" (or not worthy) and emotions such as triumph, despair, and pride. Simply put, our self-esteem is how we feel about ourselves.

WHAT GOOD IS SELF-ESTEEM?

A few decades ago, several researchers theorized that low self-esteem was associated with virtually all types of negative social behaviors and psychological disorders. They believed that those with low self-esteem tended to be under-achievers. Such people were thought likely to have psychological problems, to be dishonest, and to be involved in criminal activities.

According to those theories, people with low self-esteem were kicked out of school, got hooked on drugs, and developed other social problems. They were sociopaths, prostitutes, gangbangers, and such.

On the other hand, high self-esteem was said to be associated with all the positive characteristics that anyone could possibly aspire to. People with high self-esteem tended to be well educated, so they had higher-level jobs and made more money than other people. They were respected in the community. Successful and powerful, they usually became leaders. With few anxieties and excellent mental health, they were unlikely to become depressed.

Those with high self-esteem were believed to be generally happier with their lives than those who lacked self-esteem. For any of those fortunates, life might include graduating Phi Beta Kappa, marrying a wealthy partner, and being loved and admired by all. To top it all off, they were believed to be overly generous.

While these statements are obviously exaggerated, they are examples of how enthusiasm for self-esteem had run amok. Psychologists, educators, parents, teachers, and social scientists were expected to get on the bandwagon and announce their support of programs to raise self-esteem.

As Halvorson put it:

It's a thriving business because self-esteem is, at least in Western cultures, considered the bedrock of individual success. You can't possibly get ahead in life, the logic goes, unless you believe you are perfectly awesome.[48]

No doubt, if you were offered a choice between high self-esteem and low self-esteem, you wouldn't have to think too long before choosing the high one. It's still associated with positive things in life. But did the proponents of these theories actually prove that high self-esteem causes those better things in life? In other words, could you actually improve people's lives by increasing their self-esteem?

A deluge of research has answered that question with a flat "no."

What was wrong with that idea?

Social sciences and psychologists were dumbfounded when substantial research showed that high self-esteem doesn't cause success and low self-esteem doesn't cause failure. In fact, the researchers found that self-esteem doesn't cause anything at all.

There was no scientific evidence that any kind of self-esteem is a cause of any

---

48 Halvorson, "Forget Self-Esteem" in *Psychology Today.*

psychological condition. And further, the research clearly showed that self-esteem is a result—rather than a cause—of any individual's circumstances and behaviors.

For example, depression is not caused by low self-esteem; instead, the experiences that cause depression also cause lower self-esteem.

Successful careers are not caused by high self-esteem; instead, they result in high self-esteem.

Students don't do well in school because they have high self-esteem; instead, striving to do well in school results in high self-esteem.

Low self-esteem doesn't make teenagers abuse drugs; instead, using and abusing drugs results in low self-esteem.

Simply put, people who are successful in their endeavors tend to have high self-esteem, and people who fail to meet their goals then tend to have low self-esteem.

TRAIT AND STATE SELF-ESTEEM

Psychologists divide self-esteem into two different types, called trait and state self-esteem.

Trait self-esteem refers to how good you feel about yourself on average. When you hear someone say "self-esteem," they're generally talking about what a person feels about themselves most of the time. That's trait self-esteem.

State self-esteem refers to how you feel about yourself at this particular moment.

Your trait self-esteem probably remains pretty steady, but your state self-esteem goes up and down depending on how your day goes. If you screwed up in a particular situation, your state self-esteem drops. Even so, you can still feel good about yourself in general.

So if self-esteem doesn't cause positive or negative behavior, what is it good for?

THE EVOLUTIONARY FACTOR

When our ancient ancestors were hunter-gatherers, acceptance was critical to their very existence. Back on that African savanna, if you weren't accepted, your survival was in question, and if you were rejected, there was no longer any question—your days were numbered.

In our modern society, social rejection isn't quite as life-threatening, but that ancient adaptation is still with us. We're social creatures, and our awareness of acceptance or rejection affects our well-being. In everyday life, some troubling social situations need to be corrected, and others are of no real concern. Our sense of self-esteem evolved to help us tell the difference.

Studies have shown that all normal people care what the people they value think of them. Self-esteem goes up and down as events occur that make us feel accepted or rejected. When you believe that people like you and want to interact with you, your state self-esteem moves up. When you feel like the odd man out, your state self-esteem goes down.

Evolutionary psychologists have come up with the idea of a "sociometer theory" to explain how self-esteem works.[49] Think of an internal gauge that helps you minimize your likelihood of rejection by others. When your sociometer reads low, it motivates you to change your

49 See "Sociometer Theory" on the website PsychWiki.

behavior in order to become more acceptable. And when it reads high, it motivates you to think of the reasons why you're more acceptable.

Understanding where the self-esteem adaptation came from and how it works helps you maintain your balance. You really don't have to be extremely socially successful to feel good about yourself, but you probably do need to feel acceptable to at least a few others. Most of us just need to have a supportive group of people whose opinions we value.

8.
Feelings That Hurt

Just as feeling good is based on physical components, having your feelings hurt involves physical connections too. In fact, research has shown that the same neurotransmitters in the brain are involved in both physical pain and the pain of hurt feelings.

Pain of any kind has both sensory and emotional components. The sensation itself is physical; the response to it is emotional. Hurt feelings appear to stimulate areas of the brain that are related to physical pain and those that are related to the emotional component of pain.

Since emotions flood the brain with chemicals that have physical effects on the body, the emotional component of pain can make the physical sensations worse. (This would suggest that people who are more sensitive to physical pain are also more sensitive to hurt feelings.)

Only recently has much scientific research focused on hurt feelings, because those emotions were previously considered relatively unimportant. But anyone who says they never get their feelings hurt is most likely in denial.

FEELING THE PAIN

According to a widely used definition from the The International Association for the Study of Pain, "Pain is an unpleasant sensory and emotional experience associated with actual or potential tissue damage, or described in terms of such damage."[50]

Well, we don't need a formal definition to tell us how pain hurts. When you are in pain, you know it, without the need of definitions or explanations. It has been said that our greatest motivator is pursuing happiness, but for some people avoiding pain is the top priority.

Obviously, pain is your brain's way of alerting you that something is happening that could

---

50 See IASP Taxonomy.

threaten your well-being and maybe even your life. The pain commonly called "hurt feelings" alerts you for similar reasons. It's sounding an alarm about people who might threaten your well-being.

Over very long periods of time, evolutionary adaptations made for one purpose are often applied to other purposes.

THE EVOLUTIONARY CONNECTION

Hurt feelings are an evolutionary adaptation. For our prehistoric ancestors, hurt feelings signaled that they could be in a life-threatening situation. Feeling hurt over being rejected was a response to an actual danger, because rejection by members of your tribe meant that you might be on your way out. Without the support of the tribe, your chances of survival were about zip.

After our ancestors began to live in social groups, the biological system that produced physical pain also produced the sensation of hurt feelings.

Fortunately, rejection isn't usually life-threatening today, although it can still be painful. However, recent behavioral health studies show that not being accepted can lead to depression and that rejection is bad for our social well-being. If you've suffered rejection, as most of us have at some time in our lives, that's not a new revelation.

How to get your feelings hurt

There are lots of ways you can get your feelings hurt.

Just being avoided or ignored can do it, and so can the mere belief that you're being avoided or ignored. Of course, your perception that you're being rejected is subjective; it springs from your personality and thinking style. And as usual, people with neurotic personalities usually take a negative view of their circumstances and are especially sensitive to hurt feelings.

Countless situations can make you believe that you're being rejected. Even trivial matters can hurt—waving to someone and not having your wave returned, or not having your phone calls or emails returned. Attempting to butt into a conversation and not getting a welcoming response can cause hurt feelings, even though your attempt might have been clumsy.

Being betrayed by someone you trusted is a classic way to get your feelings hurt. Because our ancestors had to depend on each other for survival, the adaptation of hurt feelings from betrayal made sense. It's still an important topic today in plays, movies, books, and even the news media. Consider the endless stream of stories about the large and small betrayals of friends, romantic partners, business partners, politicians, and cheating spouses. On a larger scale, there is a long history of broken trust between nations.

All the ways we can get our feelings hurt stem from the sense of being rejected or avoided by a person or group with whom we'd like to have a relationship—or perhaps with whom we thought we did have a relationship.

The degree of hurt is clearly related to how much we value the relationship. For a casual connection the hurt is minimal, but if a close personal tie or a career is affected, it can be quite painful.

If your opinion of a person or group is high, you'll be hurt if they don't value you in the same way. On the other hand, there are other people that you're not too interested in, so you couldn't care less how they value you. Even though we all basically want to be liked by everyone, most of us realize that not everyone is going to be fond of everybody.

While negative criticism can hurt your feelings, one form of it can just be an attempt to get your attention. This is particularly true of male friends who playfully criticize and make jokes about each other. Usually the most popular person is the most criticized, but in a friendly way. Perhaps this mysterious behavior stems from a desire to make friends with that individual in a manly way, meaning without too much show of emotion.

YOUR HEART REACTS TO REJECTION

Feeling that you've been rejected can cause both psychological and physical reactions. According to one study, "social rejection doesn't just feel heartbreaking, it makes your heart rate drop."

Researchers at the University of Amsterdam and Leiden University had twenty-seven volunteers send in photographs of themselves and then come to a lab to participate in an experiment. [51] The volunteers were told that the study was about first impressions: their photos would be shown to students at another university who would glance at the pictures and decide whether or not they liked the person they saw. Actually, the researchers wanted to find out if social pain caused physical reactions. All of the "opinions" told to each volunteer were computer generated.

At the labs, each volunteer was hooked up for an electrocardiogram. They were asked to look at photos of actual students from another university and to guess whether the people in those photos liked them.

The researchers found that each volunteer's heart rate fell as they anticipated the other person's opinion. If the volunteer was told that the other person did not like them, the heart rate dropped more and took longer to get back to normal. And the heart rate also slowed more if the volunteer had expected a better opinion than the one they got.

---

51 From an article by Bill Hendrick on WebMD under "Health News."

The researchers concluded, "Unexpected social rejection could literally feel 'heartbreaking.'"[52]

---

52   See "IASP Taxonomy."

9.
How Relationships Work

In the chapters about feelings and self-esteem, we saw why it was essential for our tribal ancestors to associate with other humans. Their world was far too dangerous for solitary survival, so attempting to live alone was tantamount to a death sentence. Far back in prehistory, our ancestors made the adaptation for belonging to a group, and that's still important to us today.

Relationships are a central priority in most of our lives. According to Abraham Maslow's well-known hierarchy, the first two basic human needs are the physical requirements for survival and safety from harm. Our need for love (sexual or nonsexual) and acceptance from family and peer groups is third.[53] But, in fact, our need to belong is so genetically ingrained that it can overcome other concerns—even those that affect our own safety. That helps explain why children remain attached to abusive parents or adults stay in abusive relationships.

Since all individuals and circumstances are unique to some degree, every relationship is somewhat unique. We form close relationships in the hope that we'll meet each other's needs. Our more casual relationships might be based on a joint effort to reach a mutual goal. What makes any kind of relationship successful might vary in the details, but viewed from a broader perspective, some interesting similarities appear.

A MATTER OF BIOLOGY

Human beings crave social contact. It's vital to our well-being that we remain connected to others through talking, touching, and relating

---

53 In 1943, psychologist Abraham Maslow published his theory about human motivation, and he expanded it in his 1954 *Motivation and Personality.*

honestly with those we care about. "Research confirms that our need for intimacy begins at birth," observed Gary Small, a professor of psychology and aging.

> Scientists followed a group of infants who were sufficiently nourished but were not held or caressed by their mothers and found that more than half of the babies experienced developmental delays later on in life.[54]

From a biological perspective, ignoring our genetic drive for intimacy increases the amount of cortisol and other stress hormones in our systems. That heightens stress, which raises the risk for age-related diseases, including Alzheimer's, heart disease, and diabetes. Stress threatens both mental activity and physiological well-being.

On the other hand, bonding produces feelings of belonging, which increases the release of oxytocin, one of the more important "feel-good"

54 "We Are Hardwired to Be Social," by Dr. Gary Small, MD. Dr. Small also discusses the next two experiments described here.

hormones. Researchers in the new field of social neuroscience are just beginning to understand the brain circuitry that influences our need for contact. According to Small, "The neural networks in our brains are designed to keep us connected to others."

THE POWER OF JUST HOLDING HANDS

Science has also shown that certain kinds of connections are particularly beneficial. Dr. James Coan and other researchers at the University of Virginia found that simply holding someone's hand can change your brain's response to a threat.

The participants in the experiment were happily married women. When an electric shock was delivered to their big toes, how much pain they felt varied with external circumstances. They felt less pain when someone held their hands. They felt much less pain when their husbands held their hands. In each case, recordings of their brains showed significant differences in the activation of the frontal lobes, the area that's sensitive to threats.[55]

---

55 Coan is a Professor of Psychology and Director of the Virginia Affective Neuroscience Laboratory at the University of Virginia. See articles by Benedict, Johnson, and McKean.

So holding hands with a spouse can actually affect the wiring of your brain and protect your well-being. This, of course, assumes that you and your spouse are not at odds.

PHYSICAL PAIN AND EMOTIONAL PAIN

We've seen that "hurt feelings" can really hurt. A study at the University of Michigan may help explain how social rejection can actually cause physical problems. These scientists found that the same neural pathways process intense emotional and physical pain.

In this experiment, the subjects were forty people who had experienced a recent rejection by a partner. The researchers used fMRI scanners to examine their brains during two activities.

First, the volunteers were shown a picture of the partner who had rejected them. Then they were given a mild physical pain, similar to holding a very hot cup of coffee. In both cases, the same areas of the brain lit up on brain scans. Scientists had previously thought that those brain regions were only affected by physical pain, but according to Small, this study suggests that "the emotions people experience when rejected in a relationship can be comparable to physical pain."[56]

THEORIES ABOUND

Interactions with others affect our feelings and beliefs about ourselves, so our relationships contribute to our sense of who we are. Research in a field called relational self theory has found that contact with a person who reminds us of a past relationship can influence our behavior. That's because someone who resembles a significant other can change how we think about ourselves at that moment.

Substantial research has focused on intimate relationships, such as marriage, cohabitation, and dating. We have powerful genetic motivations for such bonds. In fact, the motivation for intimacy is one of our core genetic drives, which is obvious from the billions of people in the world. Other types of relationships also have some ancestral adaptations, along the lines of "you scratch my back and I'll scratch yours."

---

56 Dr. Gary Small in "We are Hardwired to Be Social."

According to several theories, psychological costs and rewards drive our decisions about relationships. We all look for rewards in our interactions with others, and we're willing to accept a certain cost for those rewards. Of course, rewards have a positive value and costs have a negative value.

A set of ideas called interdependence theory categorizes costs and rewards as emotional, social, and instrumental and by opportunity. Emotional rewards and costs are the positive or negative feelings experienced in the relationship. Social rewards and costs have to do with pleasure or displeasure about the social situations connected to the relationship. Instrumental rewards and costs are based on how well the partner handles tasks that need to be done (such as household responsibilities). Opportunity means the gains that a person gets from the relationship, or what they have to give up for it.

A similar concept, called social exchange theory, lists time, money, and effort as costs. Acceptance, support, and companionship are rewards. The people in a social exchange take responsibility for each other, and they depend on each other.

According to these theories, you calculate the worth of a relationship by subtracting the costs from the rewards. The worth of the relationship depends on whether you get a positive or negative number for your total. Positive numbers indicate that a relationship might be expected to last; negative numbers indicate that an end is probably in sight.

If the result is positive for both parties, the relationship will be stable. If it isn't, one or both of you ends up comparing what you have to possible alternatives, meaning "shopping around" for something better.

This theory isn't very different from the way we manage our economic assets. While it seems like a rather cold way of considering relationships, if you really consider your life experiences, I think you'll find this concept is right on the mark.

LEAVING RELATIONSHIPS

Why people leave unhappy relationships is easier to understand than why they leave seemingly successful ones. Although the reasons might seem obscure, the answer often lies in that balance of costs and rewards—the perceived benefits have fallen below their comparison level for alternatives. This simply means people think they can find a better relationship somewhere else.

In an effort to understand why certain relationships are better than others, researchers have studied couples from the start of their associations. Reviewing the same pairs periodically, they analyzed the features that predict later problems. In many of those developing relationships, the balance between costs and rewards changed over time. At first, the individuals were focused on rewards, but as the relationship continued, their attention turned to costs.

Often, there comes a time when the costs equal the rewards, and a relationship tends to plateau. Dissatisfaction sets in and may lead to a breakup, even after a number of years together. But in successful relationships, rewards keep on coming, the people involved accept the costs, and satisfaction tends to be on the upswing.

In an intimate relationship such as marriage, the costs are initially put on hold because they are usually overshadowed by the honeymoon phase. However, as the excitement of sexual rewards starts to decrease, the costs become more apparent. Studies show that, on average, satisfaction with intimate relations such as marriage starts to go downhill in the second year. There are two basic reasons for the decline.

One reason is a declining interest in trying to keep a partner happy. As the initial excitement wears off, the effort to be nice and responsive to each other's needs can decrease. As time goes on, the partners in an intimate relationship such as marriage tend to focus on their own interests rather than on mutual interests.

Another problem can be traced to differences between male and female genetic adaptations and life experiences (which are epigenetic influences). From an evolutionary point of view, male adaptations are rooted in the male culture, such as competition and sports.

Men are more aggressive from early childhood. That's frequently demonstrated by male two-year-olds punching each other. Soon they're looking to play with toy guns, and they become involved in sports and physical games. The result is that men are focused on action rather than on emotions.

In our modern world, men can find release for their aggression and competition by watching football and baseball games, violent movies, and combat video games on their smart phones. After retirement, men tend to revert back to their younger years by playing games such as golf and cards. Typically, men in groups rarely become involved in emotional discussions.

Females are more genetically motivated to take care of the children. In prehistoric times, they kept the campfires burning while they cooked and socialized with the other females. Generally speaking, the female life experience has been centered on emotions. They focus on getting along with social groups, and their discussions center on home life. Of course, this is changing as females increasingly take over provider roles.

In general, women come into a married union in the role of emotional manager, while men arrive with much less appreciation of the importance of emotional communication for helping the relationship survive.

Indeed, the most common desire that women express is for the couple to have good communications. For women, intimacy means talking things over, especially about the relationship itself. Men, by and large, don't relate to this approach, causing men to typically say, "I want to do things with her now, and all she wants to do is talk."

Regardless of all those built-in problems, social research suggests that the way to keep the relationship successful is to be a rewarding partner. Studies have shown that there are two key components to maintaining a successful relationship—simply to be nice and to be responsive.

Responsiveness can be defined as actively supporting and promoting the other person's well-being. All too often, misperceptions get in the way. Obviously, a person must perceive a partner's responsiveness and feel that it's actually helpful. A partner who doesn't notice a helpful response will generally retract his or her own efforts. Neither one wants to be emotionally hurt by investing more into the relationship than the other partner seems to be putting into it.

THE BALANCE

People satisfied with their close relationships feel that they are getting satisfactory rewards in return for acceptable costs. This comes about only if both people trust their partners enough to appreciate and care for them. To sustain a successful relationship, each partner must strive to be nice, responsive, and trustworthy.

10.
Conflict Arises

Individuals and groups with different interests, ideals, and goals act out of an evolutionary adaptation to compete. Our brains were shaped in prehistoric times, when competition for limited resources was intense. As they always have, people today clash when they're competing for the same resources. The problem might be a shortage or unfair distribution of territory, jobs, income, living space, or natural resources. Internationally, conflicts are also common when people feel dissatisfied with the way they are governed.

So we shouldn't be surprised that there is so much prejudice and conflict in our world. On a global scale, about one hundred million people were killed in wars, ethnic cleansing, and ideological conflicts just in the twentieth century. Competition among nations over resources, religions, and ideas is at a historic high and seems to be unending.

Prejudice and discrimination are rampant in some places, and they still lurk within even the most open and democratic societies. Intolerance based on race, nationality, ethnicity, religion, politics, sexual orientation, and other characteristics causes problems for large numbers of people.

QUARRELS AMONG GROUPS

As we've discussed, one of the six core genetic motivations is to belong to a group. Going it alone just wouldn't have worked back on the African savannas, so early humans lived in groups. Those hunter-gatherer tribes were often in conflict, and the motivations that drove them have been passed down genetically. That's visible today, with some nations at each other's throats. While legal systems, laws, and enforcement have imposed some restraint, the tribal instinct remains within us.

Researchers have found that just forming a group causes the members to believe that their group is better than others. We have a natural tendency to put others into categories, and then we instinctively start thinking that our own category is the most desirable. Joining a group initiates an "us-versus-them" attitude—and groupies of any kind are prejudiced against other groupies.

Research has also shown that groups are more competitive than individuals. Each group starts behaving in ways beneficial to itself and detrimental to others. Prejudice and conflict intensify when members think they are being unfairly judged relative to those in other camps.

Members of one group don't even have to know the members of the other group to consider them inferior or antagonistic. Genetic biases are not based on who the members are. The fact that it is a different group is enough for its members to be viewed in a negative way, no matter who they are. And even though members of different groups may be friends, that bond is often lost in the competition. (Think Democrats and Republicans, activists for various causes, and some attitudes that pass for "school spirit." Or consider national rivalries in the Olympics or team competitions, ranging from Little League to the Major League Baseball.)

It seems that the nature of the group doesn't always matter. Even some not-for-profit and charitable organizations seem benevolent until they're competing with a similar group for contributions. Religious groups show their claws to competing religions, and there have been over ten thousand recognized religious groups during recorded history. The size of the group doesn't necessarily matter either. Tribal mentality extends down to the smallest social clusters.

Religious beliefs, political systems, and ethnic backgrounds confer a sense of identity on the members of those groups. Threats to that sense of identity can evoke violent responses, ranging from angry words to physical attacks. Sometimes the need for a particular identity is strong enough to drive one group into war against another.

It's also true that leaders have used that need to gain or hold onto power.

Despite all this, conflicts can be held at least somewhat in check when cooperation rather than competition flourishes. And in modern society, joining a group still does provide certain benefits. We're inclined to trust other people who live by the same customs as we do. With similar rules of behavior, we can maintain cooperative relationships while minimizing the risks of trying to work with others. Groups also encourage an individual's transition to a "we, not me" bond, the basic principle of tribal adaptations.

ANCESTRAL ADAPTATIONS

Conflict and prejudice have genetic roots. In an attempt to put a clamp on conflict and prejudicial behavior, societies have formed a variety of laws, but these work best within with tightly controlled areas. A certain level of conflict and prejudice might be acceptable to a local population in a small area. However, as population density increases, such as in a large city, controlling conflict becomes a bigger problem. Anyone who has moved from a small town to a large city understands this fact well.

As human beings, our ancestral adaptations of conflict and prejudice are part of who we are. They are inherent in any separate groups, whether they are political, religious, or even philanthropic. Unfortunately, social scientists, evolutionary psychologists, neuroscientists, and so on have yet to come up with a practical way to even diminish this critical problem.

So far, the best we can do, at least in theory, is to make the goals of each group relatively the same. But of course, this requires that the members of one group work with and learn from members of the other group. Trying to do that sometimes adds to misunderstandings. Besides, if two groups have the same goals, it might be more effective to merge them into one. However, since merging into a larger group dilutes the power of individual members, it rarely happens.

The unfortunate conclusion is that no realistic solutions have come to light regarding large-scale group conflicts or ethnic prejudices. Consequently, world peace seems beyond reach. However, on a small scale, if different groups are educated in critical thinking, and cultural laws favor conflict control, some reduction in prejudice and hostility may be possible. To change the perspective of "us versus them" into "we" requires a common goal, such as preparation for defense against a threatening enemy. In fact, there have been a number of cases when conflict inside a country was mitigated when the leader declared a defensive war against another group or country.

So aliens from outer space threatening to destroy Earth just might turn us into "we" peoples—but even that's far from certain.

How do we control conflicts?

Although these general conclusions might not be encouraging, you can have more success controlling conflicts closer to home. You're in charge of your own responses, and that affects the people around you. Here are some of my own suggestions, along with some collected from other sources:

1. Delay your response. If your feathers are ruffled, it's best to take a moment to regroup before giving in to a gut reaction that you might regret later. Breathe deeply (in through your nose, down to your stomach and out through your mouth) to calm yourself.

2. Use cognitive appraisal. Understand that your reaction is mostly inherited and not your fault. This will give you an opportunity for self-forgiveness.

3. Avoid behaviors that escalate the conflict. Physical or verbal abuse is never acceptable. Dr. John Gottman, a leading researcher and expert on relationships, identified four additional behaviors that should be avoided during conflict: criticism (attacking the person's character), contempt (insults and nonverbal hostility, like eye rolling), stonewalling (shutting down), and defensiveness (seeing self as victim.)"

4. Show empathy. The ability to understand how the other person feels is perhaps the single most powerful means to diffuse conflict. Even when you disagree, you can show an understanding of someone else's perspectives and feelings ("I can see why you feel that way").

5. Admit that something is just your opinion. This is not a sign of weakness; rather, it shows that you are open to other perspectives.

6. Try to find a compromise. Conflicts are mostly resolved by compromise. This gives both parties an acceptable answer. Working for a win-win solution is the best goal.[57]

---

57 These tips are from Dr. John Gottman's work on conflict resolution; see articles by Joyce Marter and Marni Feuerman.

11.
When We Love

What do you mean when you use the word *love*? Do you say that you love a thing (a car, a new pair of shoes)? Do you say that you love a place (the beach, a restaurant)? Do you love a particular activity? A particular food? Do you love a pet animal? Or is love something you feel for another person?

What do you mean when you speak of "a loving person" or a "loving environment"? Are you referring to kindness, compassion, empathy, or affection?

It's pretty clear that people who use the word *love* aren't always talking about the same thing. *Love* is an "umbrella term," something that means different things to different people. Consequently, we need to agree on a definition if we're going to talk meaningfully about love.

TRYING TO DEFINE LOVE

Some 2,400 years ago, two Greek playwrights commented on the benefits of love. Euripides observed, "Love is all we have, the only way that each can help the other." Sophocles said, "One word frees us of all the weight and pain of life: That word is love."

Perhaps one hundred years after that, the philosopher Plato observed the effect of love: "At the touch of love everyone becomes a poet." His student, the philosopher-scientist Aristotle, gave us this definition, "Love is composed of a single soul inhabiting two bodies."

While we have only one word for love, the ancient Greeks were smart enough to have different words for different kinds of love. They identified four forms of love:

storge—kinship, love for those to whom you're related or feel very closely connected

philia—love for your friends

eros—romantic desire, sexual love

agape—brotherly love, desiring the highest good for another

Three of those words—*agape, philia, and storge*—are used in the New Testament of the Bible.

LOVE AS THE GOAL

In spite of those observations from the ancient Greeks, historians often trace our concept of romantic love to Europe, during or after the Middle Ages. Courtly love was said to spring from the concept of chivalry. Stories originally told to and about the nobility spoke of brave knights, exciting adventures, amazing deeds, and the honor of ladies.

Whatever it may have had to do with passion, courtly love was also considered spiritual in nature. It generally involved a nobleman secretly worshiping a noble lady from afar, winning her heart with heroic deeds, and finally consummating the often-illicit passion.

Soon those tales began to spread through the general population in the popular songs of wandering troubadours. That view of love outlasted the time period. It still thrives on TV, in movies, in novels, and in songs.

Today, romance novels make up our most popular literary genre, and they sometimes reflect that twelfth-century concept of romantic love. There are also many variations in the modern love story, including elements of suspense, fantasy, history, science fiction, eroticism, humor, religion, and the paranormal. Today these tales often focus on the heroine rather than the hero. She usually learns something or wins something. But love, however you define it, is still the goal.

Can we study love?

Those varying definitions of love make it a hard thing to study. Researchers in psychology generally divide love into two types. Our affection for those we're close to, including family and best friends, is called "companionate." Our bond with a romantic partner is called "passionate." But, of course, feelings for a romantic partner will include both types if the relationship is to last beyond its passionate beginnings.

Many people see a difference between loving someone and being in love with them. Loving can be a matter of intense liking, perhaps companionate love. The ancient Greeks might have called it *storge*, *philia*, or even *agape*. Being in love is regarded as passionate love, such as the Greek *eros*.

One type of love can mutate into another, as when best friends find themselves passionately in love. Probably the most common transformation goes in the other direction, when one partner says to the other, "I love you, but I'm not in love with you." That's almost always the end of the relationship.

Some researchers have found it easier to think about what love isn't. In a *Psychology Today* article, Deborah Anapol calls love "a force of nature." She wrote, "You cannot dictate how, when, and where love expresses itself." She adds that you can buy sex partners and even marriage partners, but you can't buy actual love. And you can't turn it on or turn it off either. Anapol concludes, "Love is its own law."[58]

Others have also defined love by contrast: it isn't hate, lust, or friendship—although close friendship usually is recognized as a form of love.

---

58 Deborah Anapol, PhD, "What Is Love, and What Isn't?"

While most people believe that love can't be defined scientifically, recent research shows that science has a lot to tell us about the subject.

Your brain in love

"Love is a drug," says Helen Fisher, an anthropologist at Rutgers University. According to Dr. Fisher, one area of the brain is "a clump of cells that makes dopamine, a natural stimulant, and sends it out to many brain regions when one is in love. These are the same regions affected when you feel the rush of cocaine."[59]

Syracuse University professor Stephanie Ortigue found that twelve different areas of the brain work together to produce that feeling of euphoria called love. Chemicals such as dopamine, oxytocin, and adrenaline are all involved, suggesting that the popular idea of love as purely a thing of the heart might be off base. Ortigue commented, "Some symptoms we sometimes feel as a manifestation of the heart may sometimes be coming from the

---

59 This quote is from "The Love Drug." Helen Fisher is also the author of *Anatomy of Love.*

brain."[60] However, she also points out that activation of the brain stimulates familiar physical sensations, such as sensations in the heart and butterflies in the stomach.

Another chemical that plays a part in passionate love is phenylethylamine, or PEA. It's a naturally occurring amphetamine. PEA and dopamine are why people in love feel so lively and upbeat. PEA can also be released during some nonromantic activities, such as skydiving and eating chocolate.

Sudden love

In the Syracuse University study previously alluded to, Stephanie Ortigue found that falling in love can be very sudden, only taking about a fifth of a second. Other research has confirmed the speediness of that reaction.

Scientists at University College in London used MRI helmets to test the responses of volunteers as they looked at photographs of strangers. When a person they found attractive seemed to be looking directly at them, a part

---

60 Stephanie Ortigue, "Falling in love only takes about a fifth of a second, research reveals."

of the volunteer's brain called the ventral striatum lit up in a matter of seconds.

So the jolt you sometimes feel when you meet the eyes of an attractive stranger is real.

But the part of your brain that lights up is called the "reward center." It responds the same way to the expectation of any sort of reward. It works on gamblers and drug addicts as well as prospective lovers. Researcher Kurt Kampe commented, "It might be just as rewarding if we meet an attractive person who is interesting or who might promote our career."[61]

So we're again faced with the question of what love really is. Apparently not every jolt in the brain indicates that we've found true love.

LOVE IS LIKE AN ADDICTION

Since love is associated with such an assortment of chemicals, it shouldn't be surprising that it can work like an addiction. After the Rutgers study, Helen Fisher commented, "It's a very powerfully wonderful addiction when things are going well and a perfectly horrible addiction when things are going poorly."[62]

61 From "Furtive Glances Spark Happy Brain Waves."
62 From "'Romantic Love Is an Addiction,' Researchers Say," by Rachael Rettner.

When a passionate relationship ends, you might experience an emotional crash, possibly serious depression. The experience can be very much like the withdrawal symptoms of addicts when they stop taking amphetamines or other stimulants. Those energizing love chemicals are no longer being released, and that can make the emotional crash even worse.

In fact, your feelings over a lost love can resemble an obsessive-compulsive disorder. Sometimes it's a matter of insufficient serotonin, a neurotransmitter that regulates mood, sleep, and learning.

"Love addiction can be tough to kick," Fisher said in a 2010 CBS interview.[63] She and her colleagues studied volunteers who had been rejected after a relationship of two years or more. They found that brain regions associated with intense cocaine addiction or cigarette addiction were still very active.

But there was good news too. As time passed, there was less brain activity in the regions associated with the attachment. The brain areas that aided decision-making and

---

63 From "Love Addiction: Tough to Kick," CBS News.

emotional regulation were becoming more active.

So if you have a tough time with withdrawal from love, spend some time actively thinking about how you're going to handle it. Recognize that you're going through a period of chemical imbalance. Remember that talking about the events is more therapeutic than burying yourself in grief. When you can evaluate your gains and losses, you're learning from the experience.

LOVE IN MEN AND WOMEN

It may seem that women are more focused on love than men are. (Consider all those romance novels.) But that's a misconception. Research on brain responses associated with passionate love find just as much activity in men as in women. Some studies indicate that men may fall more easily than women do. They may actually suffer more when those relationships don't work out.

Louann Brizendine, psychiatry professor at the University of California, San Francisco, has observed that "not only is the mature male brain more receptive to closer bonds, but it's also more sensitive to loneliness."[64]

The evolutionary background

Passionate love evolved to bring human beings together for mating. Some evolutionary psychologists have even referred to passion as a trick of nature designed to get people to start having offspring. They add that once the children are on the scene, nature doesn't care about passion anymore. Family members just need to feel attached to each other in order to raise those children.

Rutgers study author Helen Fisher said, "I think the brain circuitry for romantic love evolved millions of years ago, to enable our ancestors to focus their mating energy on just one person at a time and start that mating process." She continued, "And when you've been rejected in love, you have lost life's greatest prize, which is a mating partner."[65]

---

64 From Brizendine, "Love, Sex and the Male Brain" on CNN. Louanne Brizendine is the author of the books *The Female Brain* and *The Male Brain.*
65 Fisher, "'Romantic Love is an Addiction,' Researchers Say."

Most of our lives are longer now, and the best relationships can last well beyond the raising of a family—or sometimes skip that part altogether. In any event, loving relationships can last a lifetime.

12.
Those Gender Differences

Evolution designed men and women with the characteristics that best promoted survival and propagation of the species. Obviously, that makes men and women different in many important ways. But maybe we're not as unlike as some seem to think.

Each gender does have different experiences in life. Some of those experiences are related to genetics and some to epigenetics, and some come from our culture.

We inherited genetic predispositions from our ancestors in their hunter-gatherer world, where men were the hunters and women were the family caregivers. In addition, our life experiences (epigenetics) often support those roles. For example, boys like to fight, use guns, and be involved in sports, whereas females are more prone to play with dolls, socialize, and be interested in food preparation. These tendencies are evolved adaptations from our hunter-gather days in Africa, somewhat culturally modified.

Today, these motivating factors are being reshaped. As our culture changes, the life experiences of women and men aren't as dissimilar as they once were.

Research has shown that emotional differences between the sexes are really quite small. We are more alike than we are different in our personalities, cognitive ability, and even our sexuality. Some of the problems we have with each other might come mostly from that notion that we're incompatible. No matter what the source, real or imagined differences often make communications between the sexes difficult.

That's partly because men think in terms of things and women think in terms of people.

COMPARING THE TWO

Scientists have made substantial studies of areas that seem to differ in women and men. Based on research and measurements, they give us estimates of how the sexes vary in particular attributes. In his course on the mysteries of human behavior, Professor Mark Leary reminds us that looking at groups of research studies can be more informative than just looking at just one or two.[66] This is called meta-analysis, and it gives us a better estimate of the size of those variations between sexes. As Leary explains, meta-analysis can't tell us why the differences exist, but it can tell us what they are.

In Leary's guidebook for The Great Courses series,[67] he examines a wide range of studies of gender differences. Some of the physical measurements that turn up consistently are obvious. For example, on average, there's a big difference in how far men and women

---

66 Leary is professor of psychology and neuroscience at Duke University. See Sources for links to his course.
67 Mark Leary, *Understanding the Mysteries of Human Behavior.*

can throw various things. On average, men are stronger and faster. On average, men have greater grip strength. On the other hand, on average, women are more flexible.

However, many differences in male and female behavior are probably due to both cultural and physical differences. For example, men are physically able to perform better at some sports, but it's also true that they're encouraged to participate in sports. And while women may be innately more nurturing, they're also encouraged to be more socially adept and to fill caregiving roles.

VARIATIONS IN WIRING

As measured on psychological tests, women and men show some variations in their abilities to perform specific mental tasks. Women usually excel on memory tests and in studies designed to measure empathy and emotional intelligence. Men generally do better on tests of motor skills and spatial tasks, such as map reading. This genetic adaptation explains why men won't ask for directions. Sometimes the gap in scores is small; sometimes it is quite large.

According to recent research, there might be a genetic explanation for those contrasts between the sexes.[68] Female and male brains are wired differently—literally. In a typical male brain, many of the connections run between the front and back of each side of the brain. In women, the connections are more likely to run from side to side, from one hemisphere to the other.

These variations in the wiring of nerve connections begin to appear in adolescence, about when visible sexual differentiations of facial hair and body changes appear. At that same time, gender differences show up on psychological tests.

Various studies have given us bits of information about genders. The following observations are from Jennifer Viegas' article, "10 Gender Differences Backed Up by Science."[69]

Men are better at detecting infidelity. Men appear to be better at reading subtle vocal, visual, scent and other cues indicating their partner's fidelity, concludes a study published in the journal *Human Nature.* The downside, said

---

68 Steve Connor, "The Hardwired Difference between Male and Female Brains."
69 Viegas' article is published on Seeker Media.

co-author Paul Andrews of Virginia Commonwealth University, is that that these cues aren't always accurate, and men are more likely than women to falsely suspect cheating.

Yet another study on cheating, published in the Journal of Marital and Family Therapy, found that men are more upset by sexual infidelity, and while women are more upset by emotional infidelity. Women, it should be mentioned, outperform men when identifying emotions, according to a study in the journal Neuropsychologia.

We are evenly matched in terms of intelligence. Men tend to be larger and, as a result, tend to have bigger brains. Size, however, does not necessarily correlate with intelligence. Braininess instead relies more on neuronal connections, which we help forge when learning by experience or study.

Historically, women's IQs have lagged behind those of men by up to 5 points, but now women are surpassing men in such tests. Rex Jung, an assistant professor of neurosurgery at the University of New Mexico, has found that men tend to have more brain grey matter while women have more white matter. The differences yet again are evident, but it appears that the evolutionary battle between the sexes can, at least for now, be judged as a tie.

Women usually live longer than men. Better immunity, reduced risk for blood diseases and lower risk-taking may give women an edge on longevity. Based on Centers for Disease Control data, women tend to have a life expectancy that's 5.3 years greater than men's, but the gap is narrowing. In 1978, it was 7.8 years. The good news for men is that they tend to remain sexually active longer than women do. "Interest in sex, participation in sex and even the quality of sexual activity were higher for men than women, and this gender gap widened with age," said Stacy Tessler Lindau of the University of Chicago, who worked on a related study.

PERSONALITY SCORES

Women score higher than men on most tests of agreeableness (which means being friendly, nice, and cooperative). About half of each sex will score at about the same level on these measures, but the remaining men's scores tend to be on the low end of the scale and women's on the high end. That means that women tend to be more agreeable on average. On measures of both verbal and physical aggressiveness, men score higher.

According to Duke University professor Mark Leary, there aren't many other major differences between the sexes. "On average, men tend to be a little less considerate, a little less dependable and little less responsible than women."[70]

Although women are more emotional, that difference is not large. The sexes usually report being about equally satisfied with their lives and equally happy overall. Even so, a recent poll indicated that more women than men pray or meditate, seek spiritual enrichment, see a therapist or a mental health professional, take antidepressants, buy weight-loss products, or enroll in a diet program in order to achieve happiness. Yet despite all of these efforts, more women than men labeled themselves as pessimists after the 2008 recession. However, on that poll, a whopping 64 percent of female

70 Mark Leary, *Understanding the Mysteries of Human Behavior.*

respondents said that they felt their work and personal lives were balanced. Only 52 percent of men could make the same claim.

REGARDING SEX

One difference that does show up repeatedly is in female and male attitudes toward sex. On average, men are more casual about sex than women are, and there is a large variation in those scores. In this case, meta-analysis supports our cultural stereotype.

Women's reproductive years are more limited than men's, and women don't have the potential for producing as many children. In addition, women have a greater biological investment in caring for the children. As Leary puts it,

> From an evolutionary standpoint, if women can have fewer offspring and must invest more biologically in each child, they have to be more careful not to squander their reproductive opportunities on poor choices of mates. However, because men can potentially

father many children, they could afford to be less choosy about their mates. In fact, the men who had the most children throughout evolution may have been the least selective."[71]

These biological differences concerning fertility also help to explain why husbands are generally older than wives in every society. It seems to be the best way to produce the largest number of children, and reproduction is important to both sexes. Also, an older man has traditionally had the status and power to protect those children.

In today's society, behaviors that were prompted by evolutionary adaptations might not serve us as well as they once did. Children are more likely to survive, so having a lot of them is not everyone's first priority. Cultural and economic restrictions also don't favor producing as many offspring as possible, and our postmodern cultures are organized around monogamy and moving toward shared childcare.

---

71 Mark Leary, *Understanding the Mysteries of Human Behavior.*

Men and women are alike in one important area: both prefer kind, responsive, reasonably intelligent, and capable mates.

13.
Epigenetic Therapies

As you know, you've inherited some of your attitudes and behaviors from parents and possibly from very distant ancestors. Sometimes those legacies aren't helpful in today's world. Fortunately, there are ways to change the genetic component of behaviors that make problems for you.

Even when you've inherited genes from your biological parents, they might or might not be active in your own makeup. In this book, I've described how the activation of specific genes—or "genetic expression"—can be affected by your experiences, and even by your thoughts and feelings. In this chapter, I'll show you some specific ways to make those changes.

Since the term *epigenetics* came into use in the early 1940s, the science of epigenetics has evolved tremendously, especially in our twenty-first century. In 2008, the National Institutes of Health announced that $190 million had been earmarked for epigenetics research over the following five years. In announcing the funding, government officials noted that epigenetics has the potential to explain mechanisms of aging, human development, and the origins of cancer, heart disease, and mental illness, as well as several other conditions.

That research has expanded, and scientists have focused even more closely on the influence of epigenetic factors.

According to scientist Andrew Feinberg, MD, who directs the Center for Epigenetics in the Johns Hopkins Institute for Basic Biomedical Sciences,

Epigenetic changes have been found in the lungs of smokers and cord blood of infants prenatally exposed to smoke, writes Feinberg. He also points to epidemiologic studies showing an association between famine in Sweden,

Germany and China and shortened lifespans and schizophrenia in subsequent generations, and studies in mice and humans of nutritional deficiencies that lead to disease, an indication that epigenetic changes may occur early in life and can be heritable.[72]

In addition, Feinberg says, the modern revolution in gene sequencing has revealed many mutations in cancers that control epigenetic factors.

So researchers have discovered that epigenetic effects can occur in the womb, but also over the full course of a human life span, and they can even show up in future generations. Research also indicates that you can make use of that information at a personal level.

THE PLACEBO EFFECT

Epigenetics encourages the belief that problems caused by our behavioral genes can be fixed by our own minds. The NIH division

72 Feinberg, "To Track Environmental Impact on Genome, Don't Forget the 'Epi' in Genetics Research, Johns Hopkins Scientist Says."

of Health and Human Services includes a National Center for Complementary and Alternative Medicine. The NCCAM division reports on a wide variety of health products and practices. About forty per cent of our disposable income goes to those alternative and complementary therapies. They include acupuncture, massage therapy, spinal manipulation, *tai chi* and *qigong*, and so on. [73]

These mind-body therapies are not accepted as mainstream Western medical remedies since there is not yet enough scientific evidence of their effectiveness. Any positive results from these therapies are generally thought to be due to the placebo effect.

However, that effect is very real and can be a powerful tool.

By definition, a placebo has no therapeutic value. A placebo pill contains no drugs. But a strong belief can activate a positive result, just as though some medication were actually used. So a placebo can have considerable power.

In a study reported by health writer Dawson Church, placebos cured patients about 35

---

73 Also see Dr. Baird's article "Epigenetics and Well-Being."

percent of the time. In a different study, a placebo produced better results than Paxil or Prozac in four trials. The FDA concluded that the difference between the drugs and the placebo was "clinically insignificant."[74]

Since we know that the mind affects the body, we can put epigenetic therapies to work for us.

YOUR EPIGENETIC CONTROL

Fortunately, you can exercise epigenetic control over your genetic heritage. In order for cells to respond positively, however, they must be given the right mental intervention and perceptual thought signals. The NCCAM lists mindfulness and hypnosis among alternative therapies, and there is recent evidence of their effectiveness. See the following chapters for details on how to use those therapies to deal with specific issues.

To consider one important example, stress management is vital for happiness and health. Our program for managing stress includes these natural de-stressors: mindfulness, the

---

74 Reported by Dawson Church, *The Genie in Your Genes.*

relaxation response, and self-hypnosis. When you use those techniques, in effect you are epigenetically engineering your own cells to decrease stress hormones and increase positive neurotransmitters.

Although life itself might be stressful at times, for the most part we create our own stress. Sometimes we stress out because of mistaken beliefs that money, power, popularity, and recognition bring satisfaction and happiness. But research shows that within certain limits, that is usually not the case. These beliefs were adapted in an entirely different environment from our own. Those evolutionary adaptations in human personality and thinking style are powerful. It is still hard for us to escape those drives. However, when we understand that evolutionary heritability fosters behaviors that are out of sync with our present environment, we can change.

Research show that much of our stress comes from our imaginations. As Mark Twain is often quoted as saying, "I've had a lot of worries in my life, most of which never happened."

THE NATURE OF STRESS

Chronic stress is influenced by your environment, personality, and way of thinking. The stress hormone cortisol and the hormone DHEA (dehydroepiandrosterone) need to be kept in an appropriate balance. DHEA is our most common hormone and is associated with vital health functions such as longevity and cell repair. However, both cortisol and DHEA stem from the adrenal glands. So when the adrenals are kept busy manufacturing the higher-priority survival hormone cortisol, DHEA production falls off and our health is negatively impacted. In addition, the body's stress response initiates a vast array of chemical reactions and causes many hormones and neurotransmitters to shift in response to stressful stimuli.

If you're living in an unconscious state of mind, you worry about the past and fear the future. Living in the present does away with that problem, since your unconscious mind can only operate in the past. That means you need to stay in your conscious mind and focus on your immediate reality. This is an excellent way to eliminate the fictional factor behind chronic stress.

Changing a stressful environment depends on the options you have, but there are effective epigenetic methods for adjusting your personality and way of thinking. My chapter on mindfulness gives you specific ways to use meditation for stress reduction. Some of the interventions described there require long-term practice, but even small changes can be encouraging.

It's important to understand that you inherited many of your unhelpful behaviors. You're doing the best you can with what evolution handed down to you. Really understanding that offers powerful tools for change.

Here are a few tips:

- If you just can't believe that most worry is not productive and is usually illusionary, try this: Put your mind in a quiescent state by using the mindfulness exercise and the relaxation response.

- Perform cognitive behavioral therapy on yourself to correct your wrong (illogical) thinking. Realize that your unconscious mind, with its ancient evolutionary adaptations, is the mediator of imaginary chronic stress. Learn to focus on the present moment. (See chapter 17, "Using EBT" for self-help cognitive behavioral therapy.)

- Develop the habit of being more conscious in all circumstances.
- Distract your mind from stress by doing something entertaining and enjoyable.
- Do a good deed or carry out an act of compassion—it's the one sure way to temporarily turn off stress.

THE POWER OF ATTITUDE

Two areas of study specifically illustrate the power of your personal state of mind.

As mentioned in chapter 6, doing something good for others has been shown to increase the release of dopamine. That's why volunteering has been called the "helper's high." But some research indicates that it's important to be doing that good work for good reasons.

According to an article in *Psychology Today*,

A study of over 3000 people found that people who volunteer because they want to help others, live longer than people who don't volunteer at all. In fact, those [who serve] mainly for some sort of personal benefit live no longer than non-volunteers.[75]

So your attitude matters, not just your actions.

We see a similar positive effect from a practice called the "three blessings exercise," which works to improve your general attitude.

At the end of the day, just before going to bed, write down three things you're thankful for. They don't have to be big things; just think of something you enjoyed or a good thing that happened. These are your three blessings for the day.

Doing this exercise means that you end the day focused on three things that went well. People who complete this exercise tend to report more happiness and less depressive symptoms. Researchers have found that "the improved mood can last up to six months."[76]

---

75 Tracy P. Alloway, "Give to Feel Good."
76 See "10 Ways to Boost Dopamine and Serotonin Naturally."

You can even extend the three blessings exercise and the benefits it brings. For example, look over the blessings you recorded and think about why each of these things happened.

At this point, some people like to pick one of their three blessings to tell themselves to dream about that night. According to the Centre for Confidence and Well-Being website,[77] you'll increase your chances of having a positive dream if you take these steps:

- Give the positive event a name.
- Visualize it.
- As you go to sleep, say the name over and over, visualize it, and intend to dream about it.
- In the morning, be sure write down your positive dream.

Studies about activities such as those clearly illustrate that our thoughts and feelings are much more powerful than we often realize. You can exercise your own epigenetic interventions using the methods shown here and in the following chapters.

---

77 See "3 Blessings Exercise" on the Centre for Confidence and Well-Being website.

14.
Modifying Obesity Genes

Obesity rivals heart disease and cancer as the number-one health problem facing our society. In the United States, obesity has increased by 61 percent in just the past ten years. According to recent reports based on government data, in over twenty states, more than one-third of the population is obese. In at least six states, obesity numbers are still rising. More than one news analysis has called this an epidemic.[78]

---

78 Part of this chapter is from James D. Baird PhD, *Obesity Genes and their Epigenetic Modifiers*. In that book you'll find additional information and methods of dealing with overeating problems.

Multiple studies link obesity to such devastating diseases as cancer, coronary artery disease, and diabetes—in general, a poor quality of life. It's no surprise that more people than ever are searching for a way to lose weight permanently, get in shape, and stay fit.

Unfortunately, diets don't always work. While most diets can result in initial weight losses, most people don't maintain that weight loss over time. All too soon, the weight returns.

Dr. Michael Dansinger and colleagues of the Tufts New England Medical Center looked at results from forty-six weight-loss diet studies that covered a total of nearly twelve thousand participants. Writing in the *Annals of Internal Medicine*, they reported that the average weight loss was just 6 percent, and that within five years most dieters regained all of the weight they had lost. No diet significantly bettered that average. In fact, most diets fail in the first year, and over 90 percent fail within five years. Studies from The National Institutes of Health and other research centers verify these results.

These findings should come as no surprise, since it's common knowledge that "diets don't work." In desperation, we keep trying each new variation, hoping that one will work. But that hope is based on false assumptions, and our efforts are destined to fail.

Of course, it's not the diet that fails, but rather the dieter who fails to maintain the diet. The simple biological fact is that if we reduce our calorie intake and/or burn more calories, we will lose weight. If we continue to do this, we will weigh less, no matter what diet we're following. But while the strategy of losing weight is simple, the execution is not. Reducing calorie intake, changing food preferences, and burning more calories are all complex issues, and they're linked with prehistoric human instincts and tied to our modern culture.

THE CAUSE OF THE PROBLEM

Is there a realistic way to maintain a healthy diet, or are we doomed to just keep getting fatter? Several big issues make this a tough question to answer:

Genetics. We inherited a powerful set of "obesity" genes from the days when ancient humans lived a feast-or-famine existence. To

lose weight and keep it off, we have to acknowl-
edge the power of our genetic heritage, modify
our obesity genes, and find healthy ways to
work with our natural cravings for sweets, fats,
and calories.

Insufficient physical activity Technolog-
ical advances have substantially reduced the
necessity for physical activity. To make matters
worse, we have a built-in genetic instinct to
burn as few calories as possible.

Unhealthy eating habits. We are creatures
of habit, so trying to switch to less tasty (even
though healthier) food is destined to fail unless
we change our eating habits first.

Over the past thirty-five years, researchers have studied populations that are predis-
posed to obesity, such as the Pima Indians.
They discovered a set of genetic traits that work
against most people who are trying to adopt
healthier habits, especially while surrounded
by the temptations of today's inexpensive,
high-calorie foods.

Studies published in The Journal of Internal Medicine and the American Journal of
Clinical Nutrition identified a set of ten to

twenty individual genes—collectively called "the thrifty genotype"—that encode our food preferences and make us crave fatty, sugary foods. As soon as your body senses that fewer calories are coming in, these genes direct your body to turn up hunger sensations and turn on cravings for high-calorie foods.

It's no wonder that this kind of genetic programming eventually erodes even the strongest self-motivation, pushing most dieters to give in to temptations that lead to weight gain. On top of our genetic predisposition to obesity, for the past fifty years Americans have been living in a society that increasingly encourages us to cater to our "fat" genes. Fast-food restaurants, packaged foods, and vending machines let us find food in an instant. We now live in a hypercaloric society where our genetic predisposition for high-calorie foods can be fully satiated twenty-four hours a day, seven days a week.

HEALTHIER DIETS

It's clear that some diets are healthier than others. Studies conducted by American nutrition researcher Ancel Keys and others during the 1950s and 1960s, and expanded upon by Harvard's Walter Willett, MD, show that a diet rich in vegetables, fruits, nuts, fish, olive oil, legumes, and whole grains provides the perfect ingredients for both weight loss and longevity.

Keys's landmark Seven Countries Study examined the dietary patterns and health of people in seven different countries. He found that the world's longest-lived people were in Mediterranean areas, such as the Greek island of Crete. They stay slender by consuming most of their calories from vegetarian foods and by participating in some physical activity every day. Other major nutritional studies, such as Campbell's China Study and Wilcox's twen-ty-five-year study in Okinawa, revealed simi-lar results. The foundation of healthy weight and long life is regular physical activity, a diet rich in natural plant foods, and healthy eating habits.

THE SEVEN COUNTRIES STUDY

In the early 1950s, Ancel Keys went to the southern Italian shores to study the notion that heart disease was almost nonexistent there while Americans were experiencing unusually high rates of heart disease. Back then, heart disease was almost unheard of in southern Italy and only seemed to occur in a small upper-class subculture.

At that time, Americans believed that a large steak, a baked potato smothered in butter, and a glass of whole milk made a nutritious dinner. Keys was one of the early scientists who demonstrated that diet affects blood cholesterol and that blood cholesterol is an indicator of heart disease. This work led him to spearhead one of the greatest and most influential epidemiological studies of our time, the Seven Countries Study.

Over a five-year period, the Seven Countries Study examined thirteen thousand men aged forty to fifty-nine in Greece, Italy, Croatia, Serbia, Japan, Finland, the Netherlands, and the United States. Dietitians stationed in the homes of study subjects took samples of food for chemical analysis. During a ten-year follow-up period, they examined a range of lifestyle factors in more than ten thousand men. The results provided evidence that the percentage of calories from saturated fat was linked to increased blood cholesterol levels, resulting in an increased risk of coronary heart disease.

Those countries with diets that included the highest level of saturated fat, namely the U.S. and Finland, had the highest rates of heart disease. The heart disease rates of U.S., Finnish, and Dutch men were twice that of Italian men and four times that of the Greek and Japanese men. In addition, the study found that in these Mediterranean regions, the rates of all causes of death were among the lowest in the world.

According to the Seven Countries Study, in rural Crete in 1960, men were apparently safely consuming almost 40 percent of their calories in the form of fat, of which about 30 percent was in monounsaturated fat (a healthy fat, mostly derived from olive oil). The typical Mediterranean diet consisted of about 60 percent of natural carbohydrates (vegetarian), 30 percent fat, and 10 percent protein.

Throughout the Mediterranean, bread was (and still remains) fundamental to the diet. It was hardy, made from natural ingredients, and enjoyed without butter or margarine. A number of other studies of populations in Mediterranean countries, where diets consisted largely of foods of plant origin, also showed long life expectancies and low rates of many chronic diseases. These provided further evidence that high consumption of natural plant foods confers numerous health benefits such as lower rates of coronary heart disease, several forms of cancer, and other diseases.

Turning up their noses at plant-based foods, Americans in the 1950s were celebrating their prosperity by eating high-fat foods and lying around, while people in the Mediterranean lived with what was available from the land and what they could afford. Ironically, it was soon to be discovered that Americans were eating their way to poor health, obesity, and degenerative diseases while people in the Mediterranean were doing just the opposite.

THE MEDITERRANEAN DIET

The term "Mediterranean diet" is somewhat of a misnomer. The countries around the Mediterranean basin all have somewhat different cultures, religions, and diets. These regional diets differ from one another slightly in the amount of total fat, olive oil, wine, fish, milk, cheese, fruits, vegetables, and the type and amount of meat that they include.

Here's what's significant: all of those diets are very different from our Western diet, and the rates of coronary heart disease and cancer are lower in the Mediterranean countries than in Western countries.

The lowest death rates and longest life expectancy occur in Greece. In the '50s and '60s, Greek men on the island of Crete enjoyed the longest life expectancy in the world, and men living in other parts of Greece and southern Italy were the least likely to develop coronary heart disease. At that time, premature death from heart attack for Greek men was 90 percent lower than that of American men. This is no longer true today, as their diets have become increasingly Westernized.

More than other Mediterranean diets, the Cretan diet was rich in legumes, fruit, and healthy fats that came mostly from olive oil.

The Cretan diet contained much less meat, but supplied moderate amounts of fish and alcohol, mostly in the form of red wine. The total red meat, poultry, and fish consumed on a personal basis in southern Italy was 434 grams (15.5 ounces) per week, and in Crete it was about 371 grams (13 ounces) per week. Fish consumption in the Mediterranean region in 1960 varied from 126 grams per person per week in Crete to 519 grams in Spain and 1,057 grams (38 ounces) in Portugal. In Japan, which now enjoys the highest life expectancy in the world, fish per-capita consumption has ranged from 532 to 672 grams (19 to 24 ounces) per week over the last twenty-five years.

Specific benefits

There's considerable evidence that the Mediterranean-style diet containing olive oil is beneficial to your health.

A diet rich in complex carbohydrates and fiber, with a fat source primarily from mono-unsaturated fatty acids, lowers LDL cholesterol and is associated with a low incidence of chronic heart disease.

Epidemiological studies show that in countries that follow a Mediterranean-style diet, the incidence of colon cancer is low compared with Northern European countries.

The traditional Mediterranean-style diet has been shown to lead to lower blood pressure than typical Western diets.

Epidemiological data show a strong inverse relationship between natural carbohydrates intake and relative body weight. Due to its complex (natural) carbohydrates, the Mediterranean diet offers a lower energy content than a high-fat diet, making it more suitable for prevention of obesity.

The beneficial health effect of olive oil is due to both its high content of monounsaturated fatty acids and its high content of antioxidative substances. When substituted for saturated fatty acids, it reduces total LDL cholesterol without reducing the levels of HDL cholesterol.

The Lyon Diet Heart Study in patients recovering from heart attack showed a Mediterranean diet high in monounsaturated fatty acids, such as in olive oil, protects against chronic heart disease.

CHART— TYPICAL MEDITERRANEAN DIET

FOOD	SMALL SERVINGS	FREQUENCY
Whole grains (breads, pasta, cereals)	8 or more	Daily
Vegetables	3 or more	Daily
Fruit	2 or more	Daily
Legumes (beans, peas)	2 or more	Daily
Dairy (low or nonfat skim milk, yoghurt, cheese}	2 or more	Daily
Nuts and Seeds	1 oz. or more (1/3 cup)	Daily
Fish (fatty)	6-8 oz total	Weekly
Lean animal meats/eggs/ poultry	2 or less	Weekly optional
Olive oil	2 tablespoons	Daily (15-20% of calories)
Sweets	optional	rarely

A report published in *Circulation: The Journal of the American Heart Association* describes a four-year study in France following more than six hundred men and women who had had one heart attack. They found those consuming the traditional Mediterranean diet had a 50–70 percent lower risk of a second heart attack than those in a control group. The control group consumed a typical Western diet, where almost 34 percent of calories came from fat and almost 12 percent from saturated fat. The Mediterranean diet averaged 30 percent from fat and only 8 percent from saturated fat.

ADOPTING A HEALTHY DIET

Today, virtually all major medical organizations have recommended the Mediterranean diet. Like any other diet, it requires changing your eating habits. But without an effective plan for change, it usually fails. It's not that the diet is at fault; rather, it's the simple reality that people find it very difficult to change their genetic motivations.

Since what we prefer to eat is a learned habit, we can learn to change that habit. One way of doing that is to relocate to a healthier culture and adopt the indigenous foods. However, this is obviously not realistic for most people. So how do you change your food preferences from the typical American diet, based on animal products and excessive calories, to one based on plant foods and fewer calories, and actually enjoy the transition? When most people learn that they must eat more vegetables and cut back on animal products, there is an instant reaction of "no way." It seems totally unrealistic.

Not only can it be done—it can be accomplished with minimum effort! The Obesity Genes Diet acknowledges that it's futile to fight our obesity genes, but it is possible to change their expression. Changing genetic expression is commonly done with the epigenetic modalities, which are discussed in the next section.

As the first step, the diet meets our genetic demands for fat, sweets, and calories by substituting healthy fats (olive and canola oil), healthy sweets (fruits), and healthy calories (unlimited vegetables and fruits). The Obesity Genes Diet provides a tasty blend of foods from the diets of the world's healthiest and longest living peoples—the Mediterranean and Asian cultures. While this is a good beginning, without a change of eating habits, this excellent diet will soon fall by the wayside just as other diets have.

CHANGING YOUR EATING HABITS

For a diet to work permanently, you have to actually prefer to eat the foods that are better for you. This means modifying the expression of obesity genes that are causing the problem with the epigenetic mechanism.

Epigenetics is a relatively new but fast-expanding science that enables changing genetic expression. You can use the epigenetic mechanism to modify proteins associated with your obesity genes. We know that perceptions, beliefs, and signals from the environment can change our biology. (Consider, for example, the placebo effect.) You'll find detailed explanations of how to use epigenetics in the final chapters of this book. Also, the epigenetic modifiers presented in my earlier book, *Obesity Genes*, have proven to be effective for modifying behavior and eating preferences.

For many years, psychologists have taught self-hypnosis techniques to help patients do everything from changing lifestyle habits to overcoming pain syndromes. One study of sixty overweight women published in the *Journal of Consulting and Clinical Psychology* found that self-hypnosis combined with a healthful diet resulted in twice as much weight loss as dietary changes alone. My chapter on self-hypnosis describes easy-to-learn techniques. Weight loss by self-hypnosis can be learned in as little as five to ten minutes a day. Other epigenetic modalities, such as mindfulness, will help change the unhealthy habits in your subconscious and make the transition pleasant.

Another important cornerstone of weight loss is physical activity. Research on people who have lost significant weight and kept the weight off for a year or more shows that physical activity is necessary for any successful weight-loss program. As an additional benefit, substantial research has shown that exercise encourages the flow of "feel-good" neurotransmitters.

A suggested table of various physical activities can be found in my book *Obesity Genes*, which presents an array of options to prevent boredom and increase motivation.

15.
Mindfulness

There is one particularly powerful technique that you can use to modify the expression of your genes. It's called mindfulness, and the key idea is simply "living in the moment."

Here's a useful definition from the *Psychology Today* list of core concepts:

Mindfulness is a state of active, open attention on the present. When you're mindful, you observe your thoughts and feelings from a distance, without judging them good or bad. Instead of letting your life pass you by, mindfulness means living in the moment and awakening to experience.[79]

79 From "Mindfulness" in *Psychology Today*.

In fact, this practice is critical to developing the state of consciousness you need to deal with every issue discussed in this book. For instance, if you're focusing on managing your chronic stress, you need to get into your conscious mind to realize that more stress doesn't resolve stress.

That's because to make any change, you must first be aware of what you are thinking and feeling. In other words, you must be in your conscious mind. If you're not, you won't be able to focus on those essential changes that you want to make.

Living in the Moment

In our usual state of unconscious mind, we live in the past. That's where worry just adds more stress. When you're in your conscious mind, you are free of those past recordings that play as a voice in your head. In your conscious mind, you can look at everything without judgment or bias.

You can be creative.

The sole purpose of your unconscious mind is to defend your well-being without regard to morality or human justice. While that mode of thinking was critical in the days of our ancestors, it is inappropriate in our present environment and culture. The media constantly reminds us of the damage that prejudice and conflict cause in our world, both locally and internationally.

Mindfulness is a natural mental ability that is available to anyone, regardless of education, culture, ethnicity, or background. It's not a belief or tradition, nor is it religious or scientific. It's a practical, natural way to be aware of your life, moment by moment. With daily practice of mindfulness, you can turn off your unconscious mind and participate in life as it unfolds.

Mindfulness is not a way to run away from your problems. In fact, the opposite is true. It's a state of self-awareness—the state in which world records are set, visionary concepts are developed, and you are at your peak efficiency.

In an article published by the American Psychological Association, Daphne M. Davis, PhD, and Jeffrey A. Hayes, PhD, define mindfulness as "a moment-to-moment awareness of one's experience without judgment."[80] According to Davis and Hayes, research has shown mindfulness to be an effective method for changing certain aspects of our behavior. Some of the research they summarize supports views presented in this book.

EMOTIONAL CONTROL

We all have emotional pain in our lives, whether it's from a broken relationship, hurt feelings, or some other form of social rejection. Emotional pain is an evolutionary adaptation that alerts us that there's a threat to our social well-being. This signals us to take action to alleviate the situation. However, if you're unable to take appropriate action, you can reduce your pain and turn on self-compassion just by accepting that this is part of your genetic heritage.

Emotions, by their very nature, are impulses to take action. Since emotions motivate us, it's crucial to recognize them and to bring them into balance. Research indicates that mindfulness promotes metacognitive awareness,

---

80 From "What Are the Benefits of Mindfulness?"

meaning awareness of your own thoughts. This helps you gain calmness, clarity, and concentration, and it increases your general mental well-being.

REDUCED WORRY

The mental chatter bouncing around in your head generates worry and fear. Often, we're so trapped in thoughts of the future or the past that we forget to experience, let alone enjoy, what's happening right now. As science writer Jay Dixit put it in *Psychology Today*, "We sip coffee and think, 'This is not as good as what I had last week.' We eat a cookie and think, 'I hope I don't run out of cookies.'"[81]

Our old ancestral mind, which causes all that chatter, is connected to the past. Mindfulness keeps us in the moment and cuts out that mental noise.

Several studies have shown that mindfulness does reduce worry. It also reduces the distress called "rumination"—focusing on bleak thoughts about things that happened in the past. According to an American Psychological

---

81 From "The Art of Now: Six Steps to Living in the Moment."

Association study, both worry and rumination were reduced when twenty novice meditators participated in a ten-day intensive mindfulness meditation retreat. After the retreat, the meditation group also showed fewer signs of depression than those in a control group.

STRESS REDUCTION

Staying in your conscious mind and focusing on your immediate reality helps to eliminate imaginary elements of chronic stress. That's exactly what mindfulness helps you do.

Numerous studies connect the practice of mindfulness with stress-reduction. In one case, participants in an eight-week mindfulness-based stress-reduction group had "significantly less anxiety, depression and somatic distress compared with the control group." The results suggested that mindfulness meditation enables people to regulate their emotions and to "experience emotion selectively."[82]

Other sources have also reported specific reduction in stress as a result of mindfulness.

---

82 Davis and Hayes, "What Are the Benefits of Mindfulness?"

Two decades of published research from the University of Massachusetts Medical Center Mindfulness Based Stress Reduction (MBSR) program show that participants report lasting success. They experienced dramatic reductions in pain levels, improved ability to cope with persistent pain, better ability to relax, greater energy and enthusiasm for life, higher self-esteem, and better coping skills.[83]

SATISFACTION IN RELATIONSHIPS

Connections with other people are essential parts of our lives. Our exchanges with others influence our behavior and even our own sense of who we are.

Davis and Hayes report several studies showing that mindfulness supports satisfaction in relationships and also protects against the stress of relationship conflict. One reason is that mindfulness practitioners were better able to express themselves in social situations.

---

83 MBSR was developed by Dr. Jon Kabat-Zinn. See "What Is Mindfulness Based Stress Reduction?"

OTHER BENEFITS

Other important benefits of mindfulness include increased focus, more mental flexibility, less emotional overreaction, and increased empathy and self-compassion.

Several studies show that mindfulness can also improve your memory. The meditators on the ten-day intensive retreat mentioned above also demonstrated longer attention spans and better working memories than did the controls. In another study, a military group that participated in an eight-week mindfulness training maintained better working memory under highly stressful conditions.[84]

MINDFULNESS AND GENE EXPRESSION

In 2013, a team of researchers from Spain, France, and the University of Wisconsin reported groundbreaking news. They found changes in the expression of genes in subjects who had practiced mindfulness meditation for just one day. The meditators showed reduced levels of pro-inflammatory genes. A control group did not show those beneficial changes.

---

84 Davis and Hayes, "What Are the Benefits of Mindfulness?"

The leader of the study, Richard J. Davidson, stated, "To the best of our knowledge this is the first paper that shows rapid alterations in gene expression within subjects following mindfulness practice."[85] This outcome, against the control group, provided proof that mindfulness practice can lead to epigenetic alterations of the genome.

Past studies have shown that mindfulness-based training has beneficial effects on inflammatory disorders. These new results show how that can happen. Meditation is endorsed by the American Heart Association as an effective way to lower your risk for heart disease, which is the leading cause of death in the United States.

THE STATE OF FLOW

Mindfulness helps you to downsize your problems by defusing their intensity. Just think of the most creative and productive people whom you respect. One element they share is

---

85 Christopher Bergland, "Meditation Has the Power to Influence Your Genes." You can also read about this research in an article by Jill Sakai, "Study Reveals Gene Expression Changes with Meditation."

the ability to block out distractions and focus completely on the tasks at hand. This is a state of mind in which you lose all sense of time and space. It's called "flow."

Psychologist Mihaly Csikszentmihalyi first defined that state. He called it "the secret to happiness."[86] Flow has been defined as the creative moment when a person is completely involved in an activity for its own sake. When you're so absorbed in doing something that you lose track of everything around you, including the passage of time, you're experiencing flow. As an article in *Psychology Today* put it: "To make the most of time, lose track of it."[87]

Practicing mindfulness

The voice in your head is that of your unconscious mind unwinding a recording of your past life experiences, habits, and instincts. Mindfulness should be practiced several times a day so that the voice in your head starts to calm down.

---

86 You can see Csikszentmihalyi on Ted Talks in "Flow, the Secret to Happiness."
87 Jay Dixit, "The Art of Now: Six Steps to Living in the Moment."

One excuse that Americans often use to justify their refusal to practice mindfulness is that they can't spare the time. In reality, the practice of mindfulness for a few minutes a day will free up your mind to become more productive for more hours per day. In fact, practicing mindfulness increases your efficiency.

When you become a daily mindfulness practitioner, don't be surprised when you find that you have greater energy toward the end of the day. This is because mindfulness helps to clear out your unproductive ideas and negative emotions that sap energy and induce stress. You may well discover that after about ten days of practicing mindfulness, you function more effectively because you are no longer wasting energy due to anxiety, false perceptions, or hurt feelings.

MINDFULNESS EXERCISES

These exercises are collected and adapted from various sources, as noted. You might find several variations of each one on internet sites.

You can also find a wealth of helpful exercises in my other books:

- *Behavioral Genes: Why We Do What We Do and How to Change*
- *Obesity Genes and their Epigenetic Modifiers*
- *Happiness Genes: Unlock the Positive Potential Hidden in Your DNA*
- *The Mindful Meals Diet: A Mind/Body Plan to Develop New Healthy Eating Habits*
- *The Happiness Plan: Keys to a Fuller, More Positive Life*
- *The Modern Christian's Happiness Plan*

Eating mindfully (with a raisin). Focus on a single raisin and pick it up. Place it on the palm of your hand or hold it between your fingers. Explore it with all your senses as if you have never seen it before. Scan it; explore every part of it with your eyes as it sits on your palm or in your fingers. Turn it around. Notice the texture, the light on it, its shape, and whether it is soft, hard, coarse, or smooth.

Hold the raisin beneath your nose and carefully notice the smell of it. Bring it to one ear and squeeze it, roll it, listen for any sound coming from it. Gently place the object in your mouth and feel it on your tongue: its weight, temperature, size, and texture. Explore the sensations of it in your mouth.

When you are ready, intentionally bite into it. Slowly chew until you are conscious of the impulse to swallow. Sense the food moving down to your throat and into your esophagus on its way to your stomach. Sit with the experience, noticing any vestiges remaining on your tongue or in your mouth, including feelings of satisfaction, pleasure, or aversion.[88]

Walking mindfully. Before you start, prepare the space. Find a place where you can walk for about twelve to fourteen steps before you have to turn. Removing your shoes is good, if that's possible.

Now first notice your body as you stand in stillness. Feel the connection of the body to the ground or the floor. Become aware of your surroundings, taking in any sights, smells, tastes, sounds, or other sensations. Notice any

---

88 "3. Eating Mindfulness" from the website Living Well.

thoughts or emotions and let them be. Notice your arms by your sides, or if you prefer, hold your right hand in your left hand in front of you, or clasp your hands at your back. Notice your breath moving in and out of your body. No need to change it; just let it be.

Now shift your weight to the left leg and begin to lift your right foot up. Move it forward; place it back down on the ground. Mindfully shift the weight to the right leg and begin to lift the left foot up; move it forward, and then place it back down on the ground.

And continue with this walking . . . walking mindfully and slowly and paying attention to the sensations on the soles of your feet.

When it is time to turn, maintain the flow of mindfulness and bring your awareness to the intricate process of turning. Slowly, and with attention to each movement necessary to turn, begin to walk back to where you started.

One step at a time. Lifting, moving, placing. Lifting, moving, placing.

Notice any thoughts that arise and let them be. Return your focus to the sensation of walking. Lifting, moving, placing. Notice your breath. Continue walking, taking care to notice each intricate movement required at the turns. One step at a time.[89]

Alternate nostril breathing. An interesting fact about your nostrils is that you don't breathe through them equally all the time. Right now, you will be favoring either your left nostril or your right nostril. This is simply an exercise in doing so consciously and mindfully.

Sit with your back straight and gently seal your lips. Rest your left hand on your left thigh, palm facing upward.

Take some gentle normal breaths in this position. Relax your shoulders.

You are going to use your right thumb to close your right nostril, and your right ring finger or little finger to close your left nostril.

Start by closing your left nostril with your right ring or little finger. Inhale through your right nostril and exhale through your right nostril. Repeat five times.

---

89 "4. Walking Mindfulness" from the website Living Well.

Then release your left nostril and close your right nostril with your right thumb. Inhale and exhale gently, five times. Don't force your breath, and if you need to take a break and breathe through both nostrils, then do so. This exercise should feel refreshing and balancing— not like hard work.

The next step is alternating the breath between nostrils. So close your right nostril with your right thumb and inhale through your left nostril. Close your left nostril with your ring or little finger. Lift your thumb and exhale through your right nostril. Then inhale through the same nostril. Close your right nostril with your thumb and lift your finger to exhale through your left nostril. This is one round. Repeat three to five rounds.[90]

Some mindfulness insights

Here are some important ideas to keep in mind.

---

90 See "12. Alternate nostril breathing." You'll find this one and more mindfulness exercises at the website Living Well.

Understanding that you inherited your sometimes-unpleasant behaviors helps you achieve self-compassion.

Using your conscious mind means turning off your unconscious mind, which usually controls your life with worries and fears. As you observe your thoughts, remember that all future planning is to avoid pain or increase pleasure.

The following helpful thoughts are based on the work of Jon Kabat-Zinn:[91]

• In order to live life fully, you have to be present for it.

• To be present, it helps to purposefully bring awareness to your moments—otherwise you may miss many of them.

• You do that by paying attention to the present moment, nonjudgmentally noticing whatever is arising inwardly and outwardly.

• Understand that to rest in self-awareness is more a matter of "being" than of "doing."

---

91 See Jon Kabat-Zinn, PhD, "Suggestions for Daily Practice."

• Understand that the present moment is the only time you are truly conscious and able to feel and express positive emotions such as love, forgiveness, and compassion

16.
Self-Hypnosis

To change unhelpful habits, you can learn to bypass your conscious mind and communicate with your subconscious. Hypnotherapy has proven to be one of the most effective ways of doing that.

Hypnotherapy is a safe, effective mind/body technique that is used in mainstream medicine for many purposes, including relief of chronic pain and stress management. Think of it as meditation with suggestions.

Hypnotherapy in the form of self-hypnosis is part of the epigenetic behavior therapy (EBT) plan described in the next chapter.

That familiar hypnotic state

We all enter a hypnotic state of mind on a daily basis. For instance, you may have experienced driving down a highway and suddenly becoming aware that you've missed your turn. You weren't driving irrationally, and yet your mind was focused on something else. You actually were in a hypnotic state. In other words, whenever you're doing something that you're not consciously thinking about, you are under hypnosis. Daydreaming, reading a book, watching television, meditation, and prayer are also forms of hypnotic states.

Hypnosis is simply is method of producing a deep state of relaxation in which your subconscious is receptive to suggestions. Self-hypnosis is a self-induced form of hypnosis. Although it sometimes comes so easily that it seems mysterious, self-hypnosis works in a consistent manner that can be explained, predicted, and repeated. It's biology, not magic.

Conscious and subconscious levels

Think of your mind as having two levels. Your conscious mind is the surface, and your subconscious mind is the deep mental pool underneath. Your conscious mind is the communication center where you process thoughts and ideas. It is where you think, calculate, plan, and direct your conscious actions. Your subconscious mind controls involuntary body functions, such as digestion, breathing, heart rate, temperature, and so on. We learn at an early age to walk, eat our food with a knife and fork, ride a bike, and prefer certain foods. Once we've learned those skills, the subconscious mind remembers them.

The basic theme of the subconscious mind is self-interest. Although it wants good things for you, its judgment is sometimes faulty, and what it considers good can actually be bad. It harbors many beliefs, attitudes, and values that are untrue. That's why you can simultaneously want two contradictory things. It's the reason a person can be overweight and want to be slender at the same time.

What we consciously know to be good or bad doesn't have much influence on the subconscious level. The subconscious doesn't philosophize or think logically. It will only accept what it experiences, but that can include imagined experiences. The subconscious doesn't have to stop and think in order to respond, so it's always faster than the conscious. That's why you sometimes hear yourself saying something that you wish you hadn't.

In the conscious mind, all time is now, and there is no future or past. In your subconscious mind you are simultaneously a child, an adolescent, and the present you. This is important to know when you want to suggest change. Your subconscious doesn't appear to age. It holds the same feelings throughout your life, which could explain why trauma from an incident early in life can last a lifetime.

Your unconscious mind is your storage center. It contains all the experiences you've had since birth. It also hosts your belief system, sending you guiding messages about actions contemplated or performed. It's also responsible for your intuitive or gut feelings. The subconscious mind gives us the ability to use our imagination—and when you think about it, worry is only a negative imagination.

Putting self-hypnosis to work

The power of suggestion has a substantial influence on our behavior, beliefs, attitudes, and values. We're wired to continually monitor information coming to us through the five senses. We're always on the lookout for opportunity and on guard against threats, especially at the subconscious level. In other words, we're built to respond to suggestion.

But while we're sensitive to the power of suggestion, the primary function of our subconscious is to protect us. It will reject suggestions that it perceives to be not in our best interest. For example, if you subconsciously believe that meat is necessary for better health, then a suggestion to change your eating preference to plant foods will not be effective.

In a hypnotic state, your subconscious can accept a suggestion and consider it a reality. After the hypnotic state is ended, the suggestion can become active when it's triggered by a prearranged word or act. Those posthypnotic suggestions are among your most powerful tools for changing your behaviors.

The deeper the hypnotic state, the easier it is to implant a suggestion into your subconscious mind. The more you experience the suggestion as a reality under hypnosis, the more likely it will become a reality in your waking state.

The subconscious mind is far better at working with images than with language. This is why dreams are almost entirely visual for most of us. So it is usually easier and more effective to reach the subconscious with visual rather than verbal suggestions. Visual metaphors work well as suggestions. For example, an image of you floating peacefully and serenely on calm water suggests self-control and calmness.

Emotions are also powerful for giving yourself suggestions. Strong images or ideas about yourself and your place in the world, particularly when backed by emotions, may come true. So don't say negative things about yourself, because the power of suggestion increases their chances of coming true. That's true even when you're not in a hypnotic state.

Although this may sound oversimplified, if you think you are happy, you are more likely to be happy. Consciously keeping a half-smile can be a positive reminder of the state of mind you want to hold.

STRONG SUGGESTIONS

When you're giving yourself suggestions, keep it personal. Use "I" instead of "you," or say "I am" instead of "it is."

Use a positive form rather than a negative form. Emphasize what you are going to do rather than what you are not going to do. For example, "I prefer vegetables" is better than "I do not avoid vegetables"

Because the subconscious mind has no sense of time, you need to pay attention to the timing and tense of your suggestions. Use the words "I am" rather than "I am becoming," except when you're making suggestions targeted at a particular place, time, or event (such as when sitting down to a meal).

For visual suggestions, duration is more important than the number of repetitions. The longer you hold an image in your mind, the more it is likely to become a reality. However, ten to fifteen seconds several times a day is generally effective.

Bedtime is an excellent time to apply your suggestions. Just before going to sleep, repeat your verbal suggestions or visualize your image suggestions. This loads them into the subconscious mind, which will work with them while you are asleep.

Changes that occur as the result of suggestion will almost always feel natural and effortless. When you do see some positive changes, don't assume that they happened spontaneously rather than by suggestion. If you discontinue using your suggestion too soon, the unwanted condition is likely to return.

If you don't see at least some results by the end of three weeks, reexamine your suggestions. Look for whatever is blocking the process.

CHOOSING THE RIGHT SUGGESTIONS

Using your imagination under hypnosis can be a powerful tool for change. Suppose you are overweight and see yourself as being fat every time you look in the mirror. You desperately want to lose weight, but whenever you think about yourself, you see yourself as fat.

However many diets you go on, however much you deprive yourself of food, you still know that you are fat. That image is embedded in your mind, and it affects your perception of yourself and your behaviors. But you can change the way your own subconscious thinks of you.

Choosing the right images under hypnosis will give your subconscious mind a positive picture of how you want to be. When you use positive verbal and visual affirmations, your subconscious mind soon gets the message and starts having better ideas about a slimmer you.

Suggestion triggers. A trigger produces a reaction to a posthypnotic suggestion that was planted in your subconscious mind during a self-hypnotic state. It may be a word or a touch, a sound or a visual cue, a smell or a taste, or an emotion.

The stronger the suggestion that you implant, the more effective the trigger will be. You can also create a posthypnotic sugges-tion to respond to a positive trigger that you implanted in your own subconscious mind.

THE SELF-HYPNOSIS PROCEDURE

This practical, user-friendly exercise is used in mainstream medicine and customized here for weight loss. Practice it a few times to get some experience. For your first few self-hypnosis sessions, begin with the relaxation and deepening exercises below.

The first and most critical step for communicating with your subconscious is to enter a state of relaxation.

Relaxation response. First make yourself as comfortable as possible. You may sit or lie down. Take a breath of air, fill your lungs to a comfortable level, and focus on an object in front of you. Exhale slowly, then fill your lungs again. Slowly let your eyes go out of focus. As you continue breathing in this slow manner, start progressively relaxing the muscles throughout your entire body in an orderly sequence: first feel your feet relaxing, then feel the tension going out of the muscles in your lower legs. Relax the muscles in your lower body, progressing through your chest, arms, and neck. How long this process should take and how detailed you get is a personal choice.

Deepening techniques. Of the many techniques that deepen both relaxation and the hypnotic trance, the most popular involves counting backward. Start at five and slowly count down toward zero. Picture yourself relaxing more deeply with each number. This state is similar to going to sleep, except that you continue to be aware.

Use your imagination to go deeper. For example, imagine that you are on an elevator while counting down, that each number is a different floor level, and that you see each floor number lighting up as you move downward. The last stop is zero.

When you reach zero, tell yourself that you will feel very comfortable in a favorite place in your mind that is calm and relaxing where you can have positive experiences. This may be from your memory or your imagination. Beautiful spots in nature, such as gardens and ocean beaches, have wide appeal.

Once you have entered a hypnotic state, tell yourself to place your thumb and forefinger together (or use any word or touch that is meaningful to you). This will be your trigger, and it will remind you of how relaxed and calm you were feeling at that moment.

Every time you practice self-hypnosis and reinforce your trigger, it will become stronger and stronger in your subconscious mind.

Each time you trigger your suggestion, it will work better and be more responsive. You'll also use your trigger to initiate other posthypnotic suggestions, as described in the self-hypnosis induction script.

The following self-hypnosis induction is designed to help you lose weight. If you want to work on a different behavior, just make changes in the script to fit your desired results.

This procedure makes use of epigenetic influences shown in studies to be effective and generally applicable. Like any program to modify behavior, repetition is necessary. In other words, what you get out of it will be in proportion to what you put into it.

If you perform the exercise faithfully and purposefully, chances are good that it will work for you in changing your habits.

Self-Hypnosis for Changing Your Eating Habits

This procedure must be practiced to be effective. It should be initially practiced daily for at least five minutes, using this script or with the help of a guide instructor. Do so until your new eating habits become automatic and you have lost significant weight. Keep a journal of your practice to provide motivational support.

After a little practice, you can put a time limit on the exercise if you wish. When you enter the hypnotic state, say, "Hypnosis now for five minutes," or whatever time you want. You won't need to awaken yourself consciously by counting backwards from five to zero; your subconscious mind will automatically do that for you.

After a while, the following instructions will become familiar to you and you won't need to read them each time. Many people like to record the instructions and play them while using this procedure.

If the exercise is self-administered, speak the instructions silently to yourself. Don't speak the words in brackets, just wait the allotted time.

The suggested timing can be modified for individual preferences.

First, get yourself in a comfortable sitting position where you won't be disturbed.

Then, follow these instructions:

Close your eyes and perform this breathing exercise:

Focusing on your breathing, feel your breath coming in your left nostril and then out your right nostril, using the technique of alternating nostrils discussed in chapter 15, "Mindfulness." Breathe normally and be conscious of your breathing …

Breathe in to the count of 1 and out to the count of 2.

[2 seconds]

Breathe in to the count of 3 and out to the count of 4.

[2 seconds]

Breathe in to the count of 5 and out to the count of 6.

[2 seconds]

Breathe in to the count of 7 and out to the count of 8.

[2 seconds]

Breathe in to the count of 9 and out to the count of 10.

[2 seconds]

Say, "I will stop while you repeat this breathing exercise yourself."

[15 seconds to repeat the breathing exercise.]

Say, "Now stop your counting and concentrate on my voice while you continue your rhythmic breathing."

Verbalize the following instructions:

Invite your body to release any tension, and imagine the tension is melting away as you focus on each part of the body.

Invite your feet and ankles to relax and notice how they respond.

[2 seconds]

Imagine your legs are completely relaxed.

[2 seconds]

Now feel the tension melt out of your hips.

[2 seconds]

Feel the tension in your body dissolve as you feel your abdomen relax.

Now feel the softness flow up from your abdomen up to your chest. Relax.

[2 seconds]

Now feel your arms relax and become soft.

[2 seconds]

Finally, imagine all the tension dissolving from your neck, so that your whole body is relaxed. You are now in a deep state of relaxation.

[3 seconds]

Say, "In a moment you will count down from 5, and you will imagine your favorite place; it can be anywhere that is very beautiful and private, where you feel so relaxed and peaceful."

[3 seconds]

Say, "It could be in a beautiful garden, where you can smell the wonderful aromas wafting on the light breeze, while you feel warmed by the sun and alone with the beauty of nature."

[3 seconds]

Say, "Or it could be on a sandy beach with the blue waves breaking on the shore and the sandpipers skirting the changing water edge. And you can smell the aromas and hear the sounds of the surf and feel the sand between your toes."

[2 seconds]

Say, "Now we start to count down to get to your favorite place."

Now verbalize the countdown:

5—You relax deeper.

4—You are feeling more peaceful.

3—Feel how comfortable you are—soon you will be in your favorite place.

2—Your breathing is regular.

1—You are almost there.

0—It is so good to be here.

[2 seconds]

Say, "You are so glad to be in your favorite place where you are so relaxed and have not a care in the world. Take some time to use all your senses to fully imagine your favorite place. You may hear the sounds and smell the aromas or even imagine a favorite taste."

[5 seconds]

Say, "Listen to my voice and believe that the suggestions I am going to give you will make you slimmer and healthier, and you will enjoy life so much more. You will follow them without effort."

Verbalize the following suggestions:

Imagine the disadvantages of your present unhealthy eating and exercise habits.

[2 seconds]

See how they are making you overweight and threatening your health.

[2 seconds]

Imagine how being overweight makes you look to others.

[2 seconds]

Imagine all the energy it takes to drag around all that unhealthy fat.

[2 seconds]

You can almost feel the chest pains as your arteries become more and more plugged with animal fat.

[2 seconds]

Now imagine how being slim will make you look.

[2 seconds]

Imagine that you are feeling healthier and more energetic.

[2 sec]

Imagine being able to fit into clothes you can't wear any more.

[2 seconds]

Imagine how your relationships will get better.

[2 seconds]

Say, "You want to get rid of the unhealthy habits of eating foods that make you overweight and disease prone and that decrease your quality of life, but you are unable to break these habits. You have been unable to break your unhealthy eating habits because they are rooted in your genes and childhood environment. But you believe that you can break unhealthy eating habits by replacing them with healthy ones."

Verbalize the following suggestions:

Now imagine your present self as a dark outline of an overweight figure projected on a screen.

Now imagine your new self as a thinner white outline overlaying the dark outline.

Imagine the new white outline is becoming slimmer . . . and slimmer . . . until you cover less . . . and less of the dark outline you used to be. Whenever you think of the words "SLIM FIGURE" you will see this image in your mind's eye.

Say, "Now repeat the following affirmations, after me."

[2 seconds after each]

Verbalize these affirmations:

I can change my food preference because it is only a habit.

I can satisfy my obesity genes with healthy fats, natural sweets, and plant foods.

I will lose weight and be healthier eating mostly fruits, vegetables, and whole grains.

I will practice mindful eating, so I don't eat unhealthily or emotionally.

Overeating is an unhealthy habit and I will practice eating fewer calories.

I am not worried about being hungry because I know I can always eat as much plant food as I want.

I know that more exercise will make me healthier and more energetic.

I enjoy food, but I am not interested in eating more than I need for good health and nutrition.

I am feeling better and better as I become satisfied eating less.

I eat for good health, not to try to solve emotional problems or for comfort.

Food is not a substitute for love or attention. When I eat healthy foods and exercise more, I have more energy and feel better.

As I become thinner, I enjoy life more and more and feel alert and alive.

Say, "Now is the time to implant in your subconscious healthy eating suggestions that will remain with you after you have awakened. Repeat after me."

Verbalize the following suggestions:

If I crave unhealthy foods, I will put my hand on my stomach and think "SLIM FIGURE," imagining the thinner outline of my white figure against my old, larger dark figure.

If I feel hungry at an inappropriate time, I will place my hand on my stomach and think the words "SLIM FIGURE" and imagine my white figure outline.

Say, "Now let the image of your favorite place fade, as you come up to the surface of your mind as I count up to five."

[2 seconds]

Say, "You will feel calm and relaxed as you know you can return whenever you want."

Verbalize the count:

Zero—become aware of your breathing.

One—If you crave unhealthy animal-based food, you will place your hand over your stomach and think "SLIM FIGURE."

Two—You are coming back to the surface and will become mindful.

Three—Whenever you are thinking about food, you will imagine your white SLIM FIGURE overshadowed by your larger, former dark figure outline.

Four—Whenever you think about your favorite place, you will think about your preference for natural plant foods.

Five—Take a deep breath, inhaling deeply and exhaling, as you have returned to the surface of your mind. You are fully awake and refreshed. Take a moment to adjust to your surroundings with a brief mindful breathing exercise, counting up to 10.

[wait 4 seconds]

The more you practice this induction, the sooner you will develop the new eating and exercise habits that will make you slimmer and healthier.

DEPTH AND POWER

After you have practiced self-hypnosis a few times, you may be interested in going to a deeper level to get more power into your suggestions and affirmations. Once you have relaxed and are in your favorite place, try the techniques below and see which works best for you.

• Raise your right arm a few inches up into the air. Then, as you drop it heavily back into your lap, suggest to yourself that you are going three times deeper.

• Take in a long deep breath. As you breathe out, suggest to yourself that you are going three times deeper.

• Just say to yourself the word "deeper" two or three times, telling yourself that are going deeper each time you say the word.

• Imagine that you are walking slowly down a very long staircase. Tell yourself that each step takes you deeper into hypnosis.

• Focus on any other image that you find relaxing.

17.
Using EBT

Epigenetics is a relatively new science, and its implications are staggering. The remarkable discoveries in this field are familiar to many scientists and therapists, but they're little-known to the general public. For the most part, people still don't think about their lives in terms of epigenetics. Instead, they continue to rely on responses inherited from our prehistoric ancestors to solve their problems.

This book shows you ways to deal with those problems in a more up-to-date manner. When you put it all together, it's called epigenetic behavior therapy, or EBT.

EBT is unique in that it makes research findings available in the form of a self-help program. It brings together reliable information and proven therapies that you can use on your own. With EBT techniques, you can free yourself from the negative effects of the genetic heritage that causes problems with your emotional well-being.

THE FOUNDATIONS IN CBT

Cognitive behavior therapy, or CBT, is a mainstream therapy used by most psychologists and psychiatrists, sometimes along with medications. When the work is done with an experienced therapist, CBT is considered effective by major medical organizations, such as the Mayo Clinic. The methodology is designed to help people change faulty ways of thinking. Individuals learn to replace self-defeating negative thoughts with more effective responses.

CBT actually changes brain activity. It has been used successfully to treat disorders pertaining to mood, anxiety, personality, eating, substance abuse, sleep, stress, and psychosis.[92]

In CBT, the patient and the therapist work together to help the patient modify harmful

---

92 To read more about CBT see Gillihan in Sources.

thinking patterns, and the patient may be expected to do substantial additional work outside of sessions. This work usually involves practicing certain behaviors and word associations to help reorient thinking.

Self-help CBT

Epigenetics behavior therapy, or EBT, provides a self-help form of CBT that replaces the therapist with your own effort. By becoming more conscious, you can become more realistic in your thinking.

Most of our wrong thinking comes from our intuitive (unconscious) mind, where decisions are made based on gut reactions. Conscious thinking is critical thinking based on facts. It is "right thinking."

While the effectiveness of self-help CBT can't be clinically determined, professional methods can certainly be used on a self-help basis. In conjunction with other epigenetic therapies, self-help CBT should help you make behavioral changes on your own. This requires three important steps.

COMPONENTS OF THE **EBT** PROGRAM

This book has introduced you to the genetic roots of our behaviors (which we inherit) and the epigenetic influences on our behaviors (which are influenced by our life experiences and conscious activities). Now that we recognize and accept the causes of our behaviors, we can make use of the epigenetic therapies of mindfulness, self-hypnosis, and cognitive behavioral therapy to change them.

Here is a review of the basic components of the EBT program.

Component 1—Heritability

As described in earlier chapters, heritability is that percentage of our behaviors that can be attributed to our genetic inheritance. Evolutionary adaptations for human behavior have been passed down from our prehistoric ancestors.

Until recently, genetic and social scientists taught that genes were your destiny— something you can't do much about. In other words, we're stuck behaving like our ancient ancestors in ways that are not to our benefit. However, using the EBT program can modify your adapted behaviors to your advantage. To begin with, understanding that the cause of your behaviors is partly inherited brings self-compassion and self-forgiveness for your regretted behaviors.

Reread the chapter about the behavior you want to modify. Be sure that you thoroughly understand the heritability factor in that behavior.

Component 2—Epigenetics

Even though genes don't change, the proteins that are associated with them can be modified by epigenetic signals. In other words, genetic expression—which works through those proteins—can be turned off or on. Factors in our environment, such as diet, beliefs, perceptions, life experiences, and toxic materials all affect genetic expression. The classic example of this mechanism is the placebo effect, which converts your belief into a biological change.

Component 3—Epigenetic Therapies

Earlier chapters describe the epigenetic modalities of mindfulness and self-hypnosis for specific behaviors. Both of these are powerful tools for changing any type of undesirable behavior.

A number of important studies have shown that mindfulness meditation, in addition to behavior modification, has health benefits, including increased immune functioning.[93] The practice of mindfulness improves well-being, moderates fear, and enhances self-insight, morality, and intuition. All of these functions are associated with the brain's middle prefrontal lobe area.

UNDERSTAND THE CAUSES OF YOUR BEHAVIOR

The first step is to understand the nature of your behavior and its influencing factors. Choose a behavior that you're interested in modifying and review the specific chapter that discusses that topic. Then list the heritability and epigenetic influences that affect your behavior.

For example, perhaps you have developed an introverted personality and a sense of unworthiness. This could be because you did

---

93 Daphne M. Davis, PhD, and Jeffrey A. Hayes, PhD. "What Are the Benefits of Mindfulness?" You can read more about mindfulness and its benefits on the American Psychological Association site.

not spend your childhood in a loving environment. Understanding this will help you realize that certain problems are not your fault. Just having a better understanding of the problem can make a huge change in your thinking and start you on the path to recovery.

SELF-FORGIVENESS

After you have defined the causes of the particular behavior you want to modify, you need to understand that the problem is a matter of wrong thinking. Your problem behavior has resulted from a combination of heritability (evolutionary adaptations), epigenetics (life experiences), and culture. Understanding the source of your behavior permits self-forgiveness. It gives you the wisdom that comes with right thinking. It enables you to make the behavioral changes that will improve your life.

RELAXATION

To maintain right thinking, you must have control of your mind. That means you must be thinking consciously. You'll know that you're fully conscious when that voice in your head shuts down and your mind is calm. To calm your mind, learn to relax your body. The mindfulness procedures in this book provide a most effective way to relax. You can also use the relaxation section of the self-hypnosis chapter.

Two simpler strategies for relaxation are used in CBT. Calm breathing involves consciously slowing down the breath. Progressive muscle relaxation involves systematically tensing and relaxing different muscle groups. As with any other skill, the more you practice these relaxation strategies, the more effectively and quickly they will work.

Calm Breathing

When you're anxious about something, you might realize that you feel dizzy and lightheaded. That's because we breathe faster when we're anxious, and that makes us even more anxious.

Slow and gentle breaths can help you calm down. Here's how to do it. Breathe in through the nose, pause, and then exhale through the mouth. Pause for several seconds and then take another breath. (Pausing between breaths helps avoid hyperventilating.) For a more detailed procedure, see Chapter 16, "Self-hypnosis."

Progressive Muscle Relaxation

You can relax your body by tensing specific muscle groups and then relaxing them. This strategy can help to reduce overall tension and stress levels. The most effective method is to relax one muscle group at a time.

While sitting or lying down, start by tensing and relaxing your lower legs. Then do the same with muscles in your thighs. Do the same with your hips. Move up to your stomach, then into your chest, and finally reach your arms and hands.

This progression is just a suggestion. You can decide for yourself in what order you want to work on your muscle groups.

Right Thinking

To effectively manage undesirable emotions, you'll need to identify the cause of wrong thinking and replace it with right thinking. That means looking at yourself in terms of psychological facts rather than intuition.

To think in a healthy and helpful way, first you have to recognize what you're telling yourself. Remember that your emotions are your mind's way of alerting you to a beneficial or harmful situation. But when the perception of possible benefit or harm is mistaken, those emotions will lead you astray. The key to feeling better is to use critical thinking to change your misperceptions, and to do so you must rely more on evidence and less on gut reaction.

Most of us are in our unconscious minds most of the time, paying no attention at all to how we think. If you're not used to monitoring your thoughts, it can take some effort to develop the habit. It's important to do, because that voice inside your head has a strong impact on how you feel and behave. Just learning to pay attention can help you keep track of your thoughts and recognize how they affect you.

Once you learn to notice the thoughts that are running through your mind, pay attention to the ones that make you feel bad. They're most likely the ones that need to be corrected.

Are you feeling bad for a good reason, such as the suffering of a loved one? That's perfectly normal. Or are you feeling rejected because a friend has canceled something you planned to do together? Does it make you think that there's something wrong with you? That's probably an example of wrong thinking. Your thoughts are running to the extreme, and you need to get back to reality. Review chapter 6, "Feeling Good."

Pay attention to your emotions. Take notice of changes in what you're feeling, no matter how small the shift might be. If you start feeling upset, stop right there and pin down what you are thinking at that moment. What are you telling yourself about yourself? Is what you're thinking likely to be true? Is it even possibly true? Or is it one of those negative thoughts that so easily becomes habitual whether there's any truth in it or not? After all, maybe it's just a figment of your imagination. It could be what some therapists call a "thinking trap."[94] We all have baseless negative reactions from time to time, and we have to learn how to recognize them, control them, and change them.

_____

94 You can read more about this in "Self Help— Cognitive-Behavioural Therapy" on the site Anxiety Canada.

Speak to yourself about the problem, silently or aloud depending on your preference and circumstances. Demand evidence that a negative thought is true or even probably true. Don't be easily convinced that something that might be true is in fact a reality. Make yourself be objective.

The next step is to come up with a positive thought based on reality, a replacement thought for the negatives you've been listening to. Develop your own list of positive responses that you can use when your thinking goes negative. For example, "Okay, I've always done this, but it's a misperception and I'm going to change that now," or "That's unrealistic and I'm not falling into that trap again." Make your own collection of statements that you can use. Write them down and review them from time to time.

Facing your fears

Fear is a critical evolutionary adaptation for survival, and it plays a part in most of our behaviors. When unrealistic fears run rampant, they lead to self-destructive behaviors. That ancient brain of ours is always on the lookout for some danger to our psychological or physical well-being. Consequently, we're constantly adjusting our behavior to avoid any potential threat, no matter how unrealistic or unlikely it is. You've read about examples of this problem in the chapters about stress, emotion, feelings, self-esteem, relationships, and personality.

CBT has a method for clearing out misperceptions that cause unnecessary fear. It's called "exposure," and it involves coming to terms with your fears little by little, building up to a point where you can eliminate them completely. The most effective way to change behavioral fears is to face them gradually.

First, write a list of the things that you're afraid of. Then choose one that's blocking you from some activity or preventing you from being comfortable in a particular place. Don't try to force yourself into that fearful situation all at once. Instead, figure out a way to take small steps from fear to confidence.

Perhaps you're afraid of high places. You can face this fear gradually, possibly by looking out windows from higher and higher floors in a tall building. If you're afraid of speaking in front of a group, start by speaking before a few friends. You can adapt this technique to suit your particular situation.

18.
Why We Should Take Laughter Seriously

We all inherit an inclination to laugh at least once in a while. Some of us naturally laugh quite often. About 20–40 percent of our personalities comes from our parents, and that affects the frequency and tendency of our individual laughter.

However, our legacy of laughter goes back a lot further than our immediate family. When a behavior shows up in all of the world's cultures—as laughing does—we can be sure that it's something we've inherited from ancestors far back in time.

Laughing is genetic. That's why human babies start laughing spontaneously in the first few months of life. But that genetic inheritance isn't limited to our own species. According to studies of chimpanzees, gorillas, and orangutans, it's not just humans that laugh. Those great apes are also known to make laugh-like noises when they play together or when they're being tickled.

In fact, research indicates that laughter emerged long before humans split from the evolutionary path that led to our primate cousins, between 10m and 16m years ago.[95]

That means laughter existed before humor—long before anybody was telling jokes.

Laughter became especially important to early humans as they faced the complications of living in large groups. It strengthens social bonds, which are essential when a lot of people are trying to work together.

People today are still trying to manage the challenges of living together, and laughing together still has a role to play. In fact, people are still more likely to laugh in a group than when they're alone.

————————————

95 From Ian Sample, "Laughter Evolved in Primates 10m Years Ago."

Comedian John Cleese went to India to investigate a practice called "Laughter Yoga" for the BBC series *The Human Face*. He commented,

I'm struck by how laughter connects you to people. It's almost impossible to maintain any kind of distance or any sense of social hierarchy when you're just howling with laughter. Laughter is a force for democracy.[96]

Practical laughter

Laughter has personal benefits too. The biology of laughter includes strengthening immune cells and infection-fighting antibodies, thus improving your resistance to disease. That's why it's often referred to as "the best medicine."

Laughter triggers the release of endorphins, the body's natural feel-good chemicals that we discussed in chapter 6. Endorphins promote an overall sense of well-being and can even temporarily relieve pain. Laughter also decreases stress hormones and anxiety.

---

96 See "John Cleese Explores the Health Benefits of Laughter" by Jonathan Crow.

According to practitioners of laughter yoga, even fake or forced laughter tricks the body into releasing endorphins. Their motto is "fake it til you make it."[97]

How to Laugh

For some, frequent laughter comes naturally, and for others it might take a funny event to get them laughing. As mentioned earlier, your personal laugh quota has a lot to do with genetics. If you're not blessed with a natural inclination to laugh a lot, you'll need to work at increasing your opportunities so you can enjoy the health and well-being that laughter brings.

These laugh tips are adapted from information on a website about learning how to laugh.[98]

Smile. That should be an easy one. Your body responds to a smile on your face by releasing endorphins and serotonin. According to a report on stress relief, even a forced smile relieves stress, lowers your blood pressure, and boosts your immune system.[99]

---

97 See Jonathan Crow's article on John Cleese.
98 For longer explanations and more information, see Klare Heston's "How to Laugh" at WikiHow.
99 See "SMILE! Natural Stress Relief."

Besides, a smile is catching. Make a point of smiling at others as you go about your day. Your smile just might help others smile too.

Spend time with people who make you laugh. If you're feeling low, it won't help to hang around with others who bring you down even more. That doesn't mean you have to abandon friends who aren't funny, but do seek out some time with those who spark your sense of humor.

You can also benefit from doing something to help others get back on the lighter side. For example, when conversations focus on complaints, change the subject.

Watch movies and TV programs that you find funny. Whatever kind of comedy gets you laughing can work to improve your physical and emotional well-being.

Turn off the news. Spend some time away from programs that focus on reports of horrible events and suffering. You might even choose to be informed instead by humorous talk shows. At least, switch back and forth from dark stories to lighter and warmer ones.

Laugh at yourself. This might be the most important advice of all. We all have our awkward, even stupid, moments. Laughing at yourself can make them less disturbing to you and show others that goofs aren't all that important.

Take some time to enjoy yourself. We all tend to get caught up in everyday problems. We often find ourselves too focused on a multitude of small demands as we work toward larger goals.

If you're not doing something you enjoy every day, adjust your busy schedule to allow a smile or two. Look at or listen to a funny movie . . . program . . . record . . . photo . . . cartoon . . . whatever makes you laugh.

Your mind and body will appreciate it.

OTHER BOOKS BY JAMES D. BAIRD

Baird, James D. *Behavioral Genes: Why We Do What We Do and How to Change.* CreateSpace Independent, 2015.

Baird, James D., Ph.D. *Obesity Genes and Their Epigenetic Modifiers.* CreateSpace Independent, 2012.

Baird, James D., and Laurie Nadel. *Happiness Genes: Unlock the Positive Potential Hidden in Your DNA.* AdvantageQuest Publications, 2011.

Baird, James D., Ph.D. *The Mindful Meals Diet: A Mind/Body Plan to Develop New Healthy Eating Habits.* N.p.: IUniverse, 2007.

Baird, James D. *The Modern Christians Happiness Plan.* WinePress Pub., 2000.

Baird, James D. *The Happiness Plan: Keys to a Fuller, More Positive Life.* Liguori Publications, 1991.

ALSO SEE:
Baird, James D. "Epigenetics and Well-Being." Psych Central, 11 July 2015, psychcentral.com/blog/epigenetics-and-well-being/.

Website: www.drjamesdbaird.com/

ADDITIONAL SOURCES

"10 Ways to Boost Dopamine and Serotonin Naturally." GoodTherapy.org - Find the Right Therapist, GoodTherapy.org Therapy Blog, 12 Dec. 2017, www.goodtherapy.org/blog/10-ways-to-boost-dopamine-and-serotonin-naturally-1212177

"12. Alternate Nostril Breathing – Living Well." Living Well. . 05 Dec. 2014. www.livingwell.org. au/mindfulness- exercises-3/12-alternate-nostril-breathing/

"3 Blessings Exercise." Positive Psychology Resources, Confidence, Overview, www. centreforconfidence.co.uk/pp/techniques. php?p=c2lkPTImdGlkPTMmaWQ9MTQ

"3. Eating Mindfulness – Living Well." Living Well. . 05 Dec. 2014 www.livingwell.org.au/ mindfulness-exercises-3/3-eating-mindfulness/

"4. Walking Mindfulness – Living Well." LivingWell. Living Well, . 05 Dec. 2014. www.livingwell.org.au/ mindfulness- exercises-3/4-walking-mindfulness/

"A Super Brief and Basic Explanation of Epigenetics for Total Beginners." What Is Epigenetics?, WhatIsEpigenetics.com, 18 June 2018, www.whatisepigenetics.com/ what-is-epigenetics/

"Cultivating Mindfulness." oprah.com. 20 Nov. 2014. http://static.oprah.com/download/pdfs/ presents/2007/spa/spa_meditate_cultivate.pdf

"Endorphins: Natural Pain and Stress Fighters." MedicineNet, www.medicinenet.com/ endorphins_natural_pain_and_stress_fighters/ views.htm

"Exercise and Depression." WebMD, www.webmd. com/depression/guide/exercise-depression#1

"First Ever World Map of Happiness Produced." World Map of Happiness. PhysOrg, . 05 Dec. 2014. https://phys.org/news/2006-07-world-happiness. html

"Furtive Glances Spark Happy Brain Waves." IOL News, 'IOL', 24 Nov. 2016, www. iol.co.za/business-report/technology/ furtive-glances-spark-happy-brain-waves-75078

"Happiness Genes Located for the First Time." Opinion | Thenews.com.pk | Karachi, TheNews International, www.thenews.com.pk/ latest/116916-Happiness-genes-located-for-the- first-time

"Happiness Report: Why Does Scandinavia Always Win?" Time, time.com/ collection/guide-to-happiness/4706590/ scandinavia-world-happiness-report-nordics/

"How to Reset Your Happiness Set
Point." Psychology Today, Sussex Publishers,
21 Apr. 2013, www.psychologytoday.com/
us/blog/happiness-in-world/201304/
how-reset-your-happiness-set-point

"IASP Taxonomy." https://www.iasp-pain.org/
Education/Content.aspx?ItemNumber=1698

"Importance of Holding Hands." Parents Partner
Parenting Advice & Workshops for Parents,
Teachers & Couples, www.parentspartner.com/
importance-of-holding-hands/

"John and Julie Gottman" The Gottman Institute.
www.gottman.com/about/john-julie-gottman/

"Love Addiction: Tough to Kick." CBSNews. www.
cbsnews.com/news/love-addiction-tough-to-kick/

"Maslow's Hierarchy of Needs."
Wikipedia. https://en.wikipedia.org/wiki/
Maslow%27s_hierarchy_of_needs

"Mindfulness." Psychology Today: www.
psychologytoday.com/us/basics/mindfulness

"Paychotherapy: Cognitive Behavioral
Therapy." *NAMI: National Alliance on Mental
Illness*, www.nami.org/Learn-More/Treatment/
Psychotherapy.

"Pain and Stress: Endorphins: Natural Pain and Stress Fighters." MedicineNet, www.medicinenet. com/endorphins_natural_pain_and_stress_ fighters/views.htm

"Religion in Sweden." Sweden.se, Sweden.se, 8 Feb. 2018, sweden.se/ society/10-fundamentals-of-religion-in-sweden/

"Self Help – Cognitive-Behavioural Therapy (CBT)." AnxietyBC™. www. anxietycanada.com/help-resources/cbt/ self-help-cognitive-behavioural-therapy

"Sociometer Theory." PsychWiki. "The Set-Point Theory of Happiness." Changingminds.org, changingminds.org/explanations/emotions/ happiness/setpoint_happiness.ht

"SMILE! Natural Stress Relief." DIY Stress Relief www.diy-stress-relief.com/smile.html

"The Art of Happiness; What Is Happiness?" Psychology Today, Sussex Publishers, www. psychologytoday.com/us/basics/happiness

"The Set-Point Theory of Happiness." Changingminds.org, changingminds.org/ explanations/emotions/happiness/setpoint_ happiness.htm

"To Track Environmental Impact on Genome, Don't Forget the 'Epi' in Genetics Research." ScienceDaily, ScienceDaily, 5 Apr. 2018, www.sciencedaily.com/releases/2018/04/180405093209.htm

"What Is Love, and What Isn't?" Psychology Today www.psychologytoday.com/us/blog/love-without-limits/201111/what-is-love-and-what-isnt

"What Is Mindfulness-Based Stress Reduction?" Mindfulness Based Stress Reduction. Mindful Living Programs. www.mindfullivingprograms.com/mbsr_background.php

"World's Least Religious Nations–or the Atheist Countries....." Pakistan Defence. defence.pk/pdf/threads/worlds-least-religious-nations-or-the-atheist-countries.371353/

Alloway, Tracey. "Give to Feel Good." Psychology Today, Sussex Publishers. www.psychologytoday.com/us/blog/keep-it-in-mind/201301/give-feel-good

Anapol, Deborah. "What Is Love, and What Isn't?" Psychology Today, Sussex Publishers, 25 Nov. 2011, www.psychologytoday.com/intl/blog/love-without-limits/201111/what-is-love-and-what-isnt

Bergland, Christopher. "Meditation Has the Power
to Influence Your Genes." Psychology Today:
Health, Help, Happiness + Find a Therapist.
9 Dec. 2013. https://www.psychologytoday.
com/us/blog/the-athletes-way/201312/
meditation-has-the-power-influence-your-genes

Bernstein, Elizabeth. "Personality Research Says
Change in Major Traits Occurs Naturally." The
Wall Street Journal. Dow Jones & Company,
22 Apr. 2014. https://www.wsj.com/articles/
personality-research-says-change-in-major-traits-
occurs-naturally-1398122025

Bernstein, Elizabeth. "When Being Alone Turns
Into Loneliness, There Are Ways to Fight Back."
The Wall Street Journal. Dow Jones & Company.
4 Nov. 2014. http://www.goodforyounetwork.com/
when-being-alone-turns-into-loneliness-there-are-
ways-to-fight-back/

Bort, Ryan. "The U.S. Health Care System
Has Been Rated the Worst (by Far)
among High-Income Nations." Newsweek,
14 July 2017, www.newsweek.com/
united-states-health-care-rated-worst-637114

Brizendine, Louann. "Love, Sex and the Male
Brain." CNN. Cable News Network, 25 Mar. 2010.
. 05 Dec. 2014. http://www.cnn.com/2010/
OPINION/03/23/brizendine.male.brain/index.
html

Cacioppo, Stephanie, and John T. Cacioppo. "Do You Feel Lonely? You Are Not Alone: Lessons from Social Neuroscience." Frontiers for Young Minds, a Scientific Journal Edited by and for Kids. . 21 Nov. 2014. http://kids.frontiersin.org/articles/09/do_you_feel_lonely/

Carey, Benedict. "Holding Loved One's Hand Can Calm Jittery Neurons." The New York Times, The New York Times, 31 Jan. 2006, www.nytimes.com/2006/01/31/health/psychology/holding-loved-ones-hand-can-calm-jittery-neurons.html

Church, Dawson, Ph.D. The Genie in Your Genes: Epigenetic Medicine and the New Biology of Intention. 2nd ed. Energy Psychology, 2009. Kindle and Print.

Coan, James. "News." Virginia Affective Neuroscience Laboratory. . 05 Dec. 2014. http://affectiveneuroscience.org/news/

Connor, Steve. "The Hardwired Difference between Male and Female Brains." The Independent. Independent Digital News and Media, 3 Dec. 2013. http://www.independent.co.uk/life-style/the-hardwired-difference-between-male-and-female-brains-could-explain-why-men-are-better-at-map-reading-8978248.html

Crow, Jonathan. "John Cleese Explores the Health Benefits of Laughter." Open Culture, 29 Apr. 2015, www.openculture.com/2015/04/john-cleese-explores-the-health-benefits-of-laughter-yoga.html

Csikszentmihalyi, Mihaly. "Flow, the Secret to Happiness." Ted Talks. . 05 Dec. 2014. http://www.ted.com/talks/mihaly_csikszentmihalyi_on_flow?language=en

Darwin, Charles. The Expression of the Emotions in Man and Animals. Newburyport: Philosophical Library/Open Road, 2014. Print.

Davis, Daphne M., Ph.D., and Jeffrey A. Hayes, Ph.D. "What Are the Benefits of Mindfulness." CE Corner. American Psychological Association, July-Aug. 2012. . 5 Dec. 2014. https://www.apa.org/monitor/2012/07-08/ce-corner.aspx

Deans, Carrie, and Keith A. Maggert. "What Do You Mean, 'Epigenetic'?" Genetics, vol. 199, no. 4, 2015, pp. 887–896., doi:10.1534/genetics.114.173492.

Dixit, Jay. "The Art of Now: Six Steps to Living in the Moment." Psychology Today, Sussex Publishers, 1 Nov. 2008, www.psychologytoday.com/us/articles/200811/the-art-now-six-steps-living-in-the-moment

Ebrahim, Shah. "Epigenetics: the next Big Thing." International Journal of Epidemiology, vol. 41, no. 1, 2012, pp. 1–3., doi:10.1093/ije/dys015.

Feinberg, Andrew. "To Track Environmental Impact on Genome, Don't Forget the 'Epi' in Genetics Research." ScienceDaily, ScienceDaily, 5 Apr. 2018, www.sciencedaily.com/releases/2018/04/180405093209.htm.

Fisher, Helen. "The Love Drug." . 05 Dec. 2014. http://www.premierexhibitions.com/exhibitions/4/4/bodies-exhibition/blog/love-drug

Feuerman, Marni. "Managing vs. Resolving Conflict in Relationships: The Blueprints for Success." The Gottman Institute, The Gottman Institute 4 Min Read 215,64715 Feb. 2018, www.gottman.com/blog/managing-vs-resolving-conflict-relationships-blueprints-success/

Gillihan, Seth J. "Discovering New Options: Self-Help Cognitive Behavioral Therapy." NAMI: National Alliance on Mental Illness, 1 Nov. 2016, www.nami.org/Blogs/NAMI-Blog/November-2016/Discovering-New-Options-Self-Help-Cognitive-Behav

Halvorson, Heidi G., Ph.D. "Forget Self-Esteem." Psychology Today: Health, Help, Happiness + Find a Therapist. . 26 Oct. 2014. www.psychologytoday. com/us/blog/the-science-success/201209/ forget-self-esteem

Harari, Yuval N. Sapiens: a Brief History of Humankind. Harper Perennial, 2018.

Hendrick, Bill. "Hurt Feelings Can Hurt the Heart." MD Health News. MD, 29 Sept. 2010. www.webmd.com/balance/news/20100929/ hurt-feelings-can-hurt-the-heart#1

Heston, Klare. "How to Laugh." WikiHow, WikiHow, 28 Sept. 2018, www.wikihow.com/ Laugh

Jaynes, Julian. The Origin of Consciousness in the Breakdown of the Bicameral Mind. Houghton Mifflin Company, 2000.

Jefferson, Thomas, Adrienne Koch, and William Peden. "Letter from Thomas Jefferson to William Short." The Life and Selected Writings of Thomas Jefferson. New York, NY: Modern Library, 1972. 693. Print.

Johnson, Eric. "Holding Hands Is More Important Than You Think." The Evolution Institute, 15 May 2017, evolution-institute.org/ holding-hands-is-more-important-than-you-think/

Kabat-Zinn, Jon. "Suggestions for Daily Practice." Mindfulness. http://static.oprah.com/download/pdfs/presents/2007/spa/spa_meditate_daily.pdf

Leary, Mark. "Understanding the Mysteries of Human Behavior." English. www.thegreatcourses.com/courses/understanding-the-mysteries-of-human-behavior.html

MacGill, Markus. "What is the link between love and oxytocin?" Medical News Today, MediLexicon International, 4 Sept. 2017, www.medicalnewstoday.com/articles/275795.php

Mandal, Ananya. "Dopamine Functions." News-Medical.net, News Medical, 23 Aug. 2018, www.news-medical.net/health/Dopamine-Functions.aspx

Marter, Joyce. "10 Tips For Resolving Conflict Effectively. www.celandassociates.com/10-tips-for-resolving-conflict-effectively-written-by-joyce-marter-urban-balance/

Maslow, A. H., and Gardner Murphy. Motivation and Personality. Harper, 1954.

McIntosh, James. "What is serotonin and what does it do?." Medical News Today, MediLexicon International, 2 Feb. 2018, www.medicalnewstoday.com/kc/serotonin-facts-232248

McKean, Kaye. "Importance of Holding Hands." Parents Partner Parenting Advice & Workshops for Parents, Teachers & Couples, www.parentspartner. com/importance-of-holding-hands/

Okbay, Aysu, et al. "Genetic Variants Associated with Subjective Well-Being, Depressive Symptoms, and Neuroticism Identified through Genome-Wide Analyses." Nature News, Nature Publishing Group, 18 Apr. 2016, www.nature.com/articles/ng.3552

Ortigue, Stephanie. "Falling in Love Only Takes about a Fifth of a Second, Research Reveals." ScienceDaily, www.sciencedaily.com/ releases/2010/10/101022184957.htm

Paddock, Catharine. "Does a 'happiness gene' exist?" Medical News Today, 7 Sept. 2016, www. medicalnewstoday.com/articles/309537.php

Puff, Robert. "Your Set Point for Happiness." Psychology Today, Sussex Publishers, 8 Sept. 2017, www.psychologytoday. com/us/blog/meditation-modern-life/201709/ your-set-point-happiness

Rettner, Rachael. "'Romantic Love Is an Addiction,' Researchers Say." LiveScience. TechMedia Network, 06 July 2010. www.livescience. com/6695-romantic-love-addiction-researchers. html

Roman, Kaia. "How to Trigger the Brain Chemicals That Make You Happy." Medium, 13 July 2017, medium.com/thrive-global/the-brain-chemicals-that-make-you-happy-and-how-to-trigger-them-caa5268eb2c

Sachs, Jeffrey D. "Chapter 7 America's Health Crisis and the Easterlin Paradox." World Happiness Report 2018, pp. 146–159.

Sachs, Jeffrey D. "Chapter 7 Restoring America's Happiness." World Happiness Report 2017, pp. 178–184.

Sakai, Jill. "Study Reveals Gene Expression Changes with Meditation." The University of WIsconsin News, 4 Dec. 2013. . 26 Oct. 2014. www.news.wisc.edu/22370

Salamon, Maureen. "11 Interesting Effects of Oxytocin." LiveScience, Purch, 3 Dec. 2010, www.livescience.com/35219-11-effects-of-oxytocin.html.

Sample, Ian. Our primate ancestors have been laughing for 10m years The Guardian, Guardian News and Media, 4 June 2009, www.theguardian.com/science/2009/jun/04/laughter-primates-apes-evolution-tickling

Shimer, David. "Yale's Most Popular Class Ever: Happiness." The New York Times, The New York Times, 26 Jan. 2018, www.nytimes.com/2018/01/26/nyregion/at-yale-class-on-happiness-draws-huge-crowd-laurie-santos.html

Small, Gary, M.D. "We Are Hardwired to Be Social." www.newsmax.com/health/garysmallmd/socialization-alzheimers-stress-cortisone/2014/05/30/id/574304/

Smith, Emily Esfahani. "Meaning Is Healthier Than Happiness." The Atlantic. Atlantic Media Company, 01 Aug. 2013. www.theatlantic.com/health/archive/2013/08/meaning-is-healthier-than-happiness/278250/

Viegas, Jennifer. "10 Gender Differences Backed Up by Science." DNews. 28 May 2013. www.seeker.com/10-gender-differences-backed-up-by-science-1767567133.html

Washington, George. "Washington's Farewell Address 1796." The Avalon Prject. avalon.law.yale.edu/18th_century/washing.asp

Watters, Ethan. "DNA Is Not Destiny: The New Science of Epigenetics." Discover Magazine. 22 Nov. 2006. www.discovermagazine.com/2006/nov/cover

World Happiness Report. "World Happiness Report 2018." 14 Mar. 2018, worldhappiness.report/ed/2018/

Wright, Robert. NonZero: the Logic of Human Destiny. Vintage Books, 2001.

Zak, Paul J. "The Top 10 Ways to Boost Good Feelings." Psychology Today, Sussex Publishers, 7 Nov. 2013, www.psychologytoday.com/us/blog/the-moral-molecule/201311/the-top-10-ways-boost-good-feelings

See www.drjamesdbaird.com for information on the author and his works. This site includes a blog discussing further advances in behavioral genetics, epigenetics, and means for behavioral modification. You'll also find comments and questions and answers on the blog.

About the Author

James D. Baird is a graduate engineer and researcher who has studied genetics and common behaviors for more than 20 years. He has researched the field of behavioral epigenetics, which holds the promise of modifying behavior by changing gene expression. Baird holds a PhD in natural health and is the author of five books on human behavior. He has won many awards and appeared on major TV stations and dozens of radio shows.

Dr. Baird's books include:

The Happiness Plan (1990)

Modern Christian Happiness Plan (1999).

Mindful Meals Diet (2007)

Happiness Genes (2010)

Obesity Genes (2012)

Behavioral Genes (2015)

Epigenetics and Genetic Happiness (2019)

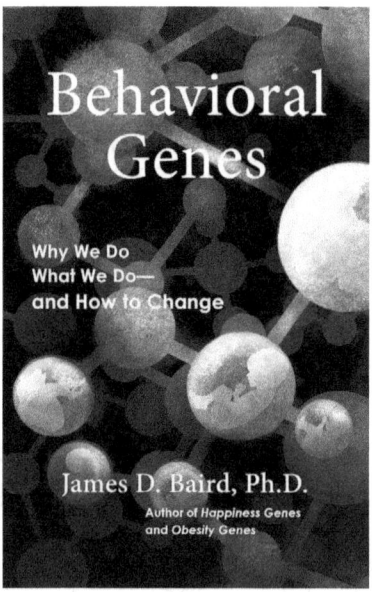

BEHAVIORAL GENES:
WHY WE DO WHAT WE DO AND HOW TO CHANGE

We humans behave in mysterious ways. We fight for seemingly no reason, eat too much, make our selves and others unhappy, and behave in all sorts of self-defeating ways. Why do we do what we do? And how can we change our behavior for the better?

The root cause of our mysterious behaviors is that 30-50% of what we do is driven by genes inherited from our prehistoric ancestors.

But we needn't be trapped by those genes or those behaviors! The new science of epigenetics proves that genes are not necessarily destiny. These exciting new discoveries are already being used by medical organizations, and Dr. Baird shows us how we can use them in everyday life.

Behavioral Genes offers specific ways to move beyond your ancient "caveman mind" and actually change your biology. It examines the sources of stress, aggression, conflict, overeating, hurt feelings, mistaken perceptions, and unhappiness. It brings understanding to core behaviors such as personality, happiness, relationships, love, and differences between the sexes.

Drawing on the proven therapies of meditation, mindfulness, cognitive therapy, and epigenetic behavioral therapy, *Behavioral Genes* maps out a positive new way to increase well-being.

Behavioral Genes cover by Tyler Baird
http://tylerbaird.net/

Happiness Genes
Unlock the Positive Potential Hidden in Your DNA

Happiness is at your fingertips, or rather sitting in your DNA, right now! The new science of epigenetics reveals there are reserves of natural happiness within your DNA that can be controlled by you, by your emotions, beliefs and behavioral choices.

Happiness Genes, written with Laurie Nadel, Ph.D., examines the nature and source of happiness from ancient times to the present. It presents research in biology and epigenetics that show DNA contains genes for natural happiness and your ultimate well-being.

Happiness Genes proves a definitive link between science and spirituality. It shows how you are biologically wired for natural happiness. It offers a 28-Day Natural Happiness Program.

> "Happiness Genes is...like reading an instruction manual that tells you how to reprogram your genes for happy life. It's time for new thinking. This book is your wake-up call."
> —**Bruce Lipton**, PhD, *New York Times* best-selling author of *The Biology of Belief*

> "One of the most universal and enduring spiritual teachings is that happiness is inside, not outside, and that it is our birthright. Breakthroughs in the science of epigenetics affirms this wisdom. If you are seeking greater fulfillment and happiness in your life, do not deny yourself this important book."
> —**Larry Dossey**, MD, author of *The Power of Premonitions* and *Healing Words*

> "Combining recent studies showing the epigenetic effects of changing our beliefs and behaviors with the wisdom of ancient spiritual traditions, *Happiness Genes* is a wide-ranging survey of the science of happiness. It asks the reader provocative

questions, like considering how our behaviors might be affecting the genetic expression of our children. Baird and Nadel show you how to inventory your 'happiness assets,' those relationships, possessions and situations that contribute to your happiness, and then treat them like the valuable resource that they are. Wise and practical, the book is peppered with self-reflection exercises, and ends with a 28-day program that guides you in cleaning up everything in your life that stands between you and happiness. To understand the links between what happiness looks like at the molecular level in our cells, the happiness of entire nations, and how you can take lessons from everything in between to raise your own happiness level, this book is an excellent guide."

—**Dawson Church**, best-selling author of *The Genie in Your Genes*

"An important book to read and re-read."

—**Edgar Mitchell**, PhD, Apollo astronaut and author of *The Way of the Explorer*

"A positive, enlightening guide for any general collection."

—*Midwest Book Review*

Obesity Genes
and their Epigenetic Modifiers

You can use genetic engineering with your mind to improve your health and your appearance. Recent research has discovered that the mechanism of epigenetics can modify genes by affecting their proteins with environmental signals.

Obesity Genes and their Epigenetic Modifiers explains the genetic facts behind why typical diets don't work. This book uniquely provides epigenetic modifiers to change the expression of your obesity genes. These known mental exercises are used in many of our largest medical organizations, and here they are customized for healthy weight loss.

Our obesity genes were designed by nature for our ancestors' survival in a limited food environment. As our environment evolved, our natural food became engineered and plentiful, feeding genetic imperatives. This set the stage for the present escalating obesity trend.

Since what we prefer to eat is a habit, this program is designed to help you develop the eating habits of the world's healthiest and longest lived peoples. *Obesity Genes* contains a mental exercise program that can provide the environmental signals to epigenetically modify or "turn off" our obesity genes. These mental acts are specially modified to be done anywhere in less than 15 minutes/day.

"Obesity Genes and Their Epigenetic Modifiers tackles head-on a pervasive conundrum—how can people who have obesity programmed into their very genetic makeup improve their weight and their health? Millennia of natural evolution have favored the survival of human beings whose bodies stored as much fat as possible to sustain them through lean times. Fad or gimmick diets can have little to no effect against such powerful Darwinian forces. However, recent discoveries of the process of epigentics indicate that humans are able to counteract the impulses of their very genes through the proper mental exercises, as well as improving their diet (by slightly reducing calories and choosing more nutritious foods), and engaging in regular physical exercise.

"*Obesity Genes and Their Epigenetic Modifiers* shows the way, by revealing the flaws in popular "weight loss" programs that can prompt a yo-yo effect, studying the world's longest-lived peoples, promoting "mindful eating" (being aware of and enjoying one's food choices, rather than zoning out or being distracted), offering guidelines for incorporating vigorous exercise into one's routine, and even laying out a sample day-by-day menu for four weeks. *Obesity Genes and Their Epigenetic Modifiers* is a solid guide to countering obesity on both physical and mental fronts, and worthy of extended consideration especially by anyone who has struggled with diets in the past."

—*Midwest Book Review*

www.ingramcontent.com/pod-product-compliance
Lightning Source LLC
Chambersburg PA
CBHW071252220526
45468CB00001B/92